SCIENTISTS AS PROPHETS

SCIENTISTS AS PROPHETS

A Rhetorical Genealogy

Lynda Walsh

OXFORD
UNIVERSITY PRESS

OXFORD
UNIVERSITY PRESS

Oxford University Press is a department of the University of Oxford.
It furthers the University's objective of excellence in research, scholarship, and education by publishing worldwide.

Oxford New York
Auckland Cape Town Dar es Salaam Hong Kong Karachi
Kuala Lumpur Madrid Melbourne Mexico City Nairobi
New Delhi Shanghai Taipei Toronto

With offices in
Argentina Austria Brazil Chile Czech Republic France Greece
Guatemala Hungary Italy Japan Poland Portugal Singapore
South Korea Switzerland Thailand Turkey Ukraine Vietnam

Published in the United States of America by
Oxford University Press
198 Madison Avenue, New York, NY 10016

Library of Congress Cataloging-in-Publication Data
Walsh, Lynda, 1971–
Scientists as prophets : a rhetorical genealogy / Lynda Walsh.
pages cm
Includes bibliographical references (pages) and index.
ISBN 978–0–19–985709–8 (hardcover : alk. paper)—ISBN 978–0–19–985711–1 (pbk. : alk. paper)
1. Science—Social aspects—Forecasting. 2. Scientists—Moral and ethical aspects.
I. Title.
Q175.5.W35 2013
303.48'3—dc23
2012038843

CONTENTS

PREFACE

The first recorded proto-scientific experiment with a control, as far as I know, was initiated by the ancient Hebrew prophet Daniel around 600 B.C. Daniel's people had been taken captive by Nebuchadnezzar, the king of Babylon, who chose Daniel and some of his companions to serve him based on their looks and intelligence. To show his favor, the king fed the Jewish youths from his own table, but he routinely ate meat and wine that had been sacrificed to the Babylonian gods; this was a dietary practice forbidden to Jews by Yahweh. When the steward in charge of the youths noticed them abstaining, he implored Daniel to eat, arguing that if the young men turned "haggard," the king would have his head. Daniel said to the man (I paraphrase), "I'll make you a deal: give us 10 more days just eating the vegetables and water, and at the end of those days, compare us to the young men who have been eating all of that plus the meat and wine. If you find that we look haggard, we'll eat as you direct us from then on. But if we seem the same as the meat-eaters, or better even, then you must allow us to observe our traditional diets." At the end of the ten-day trial, Daniel and his companions appeared "fatter" than the Babylonian men, so they were allowed to maintain their diet; thus, the story concluded, God provided for his people even in captivity (Dan. 1:1–21).

My mother first pointed this story out to me when I was a child. I think it caught her imagination because she was a practicing chemist at that time with an interest in nutrition. But she didn't, and I didn't, dig into the story to look for deeper links between prophetic and scientific practice. It wasn't until I was in graduate school at the University of Texas that I started remarking the similarities in the two discourses. I attended a lecture by the biologist Dr. Martin Poenie on the historical development of science's cultural authority. He pointed out that beliefs we had once founded on religious bases were now founded, or overturned, on the basis of public arguments made by scientists. I don't remember exactly how

Dr. Poenie phrased this thesis, but Joseph Ben-David has articulated a closely related version:

> [S]cience has a decisive influence on all the cognitive structures whereby man orients himself in the universe, in nature, and in society. Its impact began with the overthrow of biblical and classical astronomy, which used to place the earth in the center of the universe, and it continued with the overthrow of the views on the creation of the earth and man. And science's impact is perhaps strongest today as a result of its increasing mastery over illness and the consequent elimination thereby of one of the sources of permanent anxiety and hope around which so many religious practices centered.[1]

I reflected on the ways in which I myself paid obeisance to scientists: they told me sunburns caused skin cancer, and I dutifully started applying sunblock; they told me a low-fat diet encouraged weight loss, and I ate low-fat baked goods; they told me that a low-carb diet encouraged weight loss, and I gleefully starting eating olive oil again but cut out pasta. I did this reflexively, even religiously—and I, of all people, should have known better. I grew up completely surrounded by research scientists and knew for a fact that they were just regular people with foibles and biases, not superhuman geniuses. But still…

I reflected then on the scientific voices that were the loudest in the media: at the time, they were Carl Sagan, Stephen Hawking, and Stephen Jay Gould's. I noticed then how frequently these science celebrities mentioned "God" even though they were avowed atheists or agnostics. I noticed how similar their speech patterns were to the biblical prophecy I had read in Daniel and heard in church as a child. Something was going on here: something about the immense cultural authority of scientists was linked to the way they imitated prophets when they spoke to the public. That gestalt was the genesis of this project.

I think I need to say that I am not arguing, nor do I believe, that all scientists profess a sort of religious faith in science. First of all, in this book I am not examining scientists in their daily interactions with each other. Rather, I am considering how they address the public when we call on them to do so through media outlets or our elected representatives. So, I am treating a special role called the "science adviser" rather than the "scientist" proper. Even with this distinction made, I need to stress that I do not think all science advisers are evangelical, out to convert everyone to rational materialism. Versions of that argument have been made recently by David Tietge, Brian Appleyard, and in a more restricted sense, by Thomas Lessl in his study of the "priestly" *ethos* of evolutionists like Richard Dawkins. While I respect this work, I have drawn a different conclusion from the data. As I have delved into the prophetic ethos, what has impressed me is how adaptable it is. In a nutshell, the prophetic ethos *manufactures political certainty in times of crisis*. This role can be played by anyone who has privileged access to knowledge and can use that leverage to start a dialogue about public values. Scientists frequently step up into the "bully pulpit" of this ancient ethos, but so do social and environmental reformers, politicians, preachers, and sometimes even professors. I am interested in the prophetic ethos of science advisers because I was brought up right at the crossroads of these two discourses, with one in each ear and marveling at how similar they sounded at times.

If those were the roots of this project, I can mark its sprouting 13 years ago with a paper I wrote in Davida Charney's *Rhetoric of Science* seminar on the prophetic discourse of Sagan, Gould, and Hawking. That effort branched into a study of the Delphic oracle for Rosa Eberly's *Ancient Rhetorics* seminar and kept growing from there. I am indebted to these women for their indulgence of my scholarly obsession, and I hope they see in this book after all this time both the fruit of their guiding wisdom and my abiding gratitude for it and for them.

Those thank-yous begin a necessarily long list of them because no project of this magnitude grows to fruition without a great deal of tending on all sides: first, to my family for many discussions over these long years; especially to my husband Patrick, whose sanity, if not his very safety, was tested by teetering stacks of books all over the house. I don't know how he managed to listen to my thousandth anecdote about Oppenheimer without falling asleep in front of me, but I'm so grateful to him for that I can't say. Next, thank you to my colleagues and friends who have sustained me intellectually during the research and writing of this book, particularly Jane Detweiler, Cheryll Glotfelty, Eric Rasmussen, David Fenimore, Erin James, James Mardock, Anupama Mohan, Justin Gifford, Chris Coake, Stephanie Lauer, Amy Fitch, Fiona Murphy, and Steve Gehrke. A shout-out to Ben James, too, for technical assistance. I am indebted to Dan Seward, Cathy Chaput, Aimée Kendall Roundtree, Harold Delaney, and Thomas Lessl for reading chapter drafts. By a freak of research nature, I read Lessl's work *after* I had completed the greater part of my analysis. While I regret this accident because of how much time I wasted reinventing substantial parts of Lessl's wheel, yet I find my independent replication of many of his results to be heartening. Our accounts have key differences, as I mentioned above, but in the end I count the similarities between them to be more telling.

I have had outstanding help from graduate student research assistants. Kenny Walker brought me most of my primary sources on Rachel Carson from the Beinecke Library at Yale and the National Archives. Dave Johnson and Jamie Albright helped by reading source material and by discussing research hypotheses with me. Andy Ross diligently checked citations. All of this assistance helped sharpen the project. I'd also like to thank the students who critiqued parts of the book in seminars I taught on Renaissance rhetoric, writing professionalization, and style.

This project would have been utterly impossible without the expert assistance of numerous librarians and archivists. I would particularly like to thank Nicola Court at the Royal Society library in London; the staff in the Manuscript Reading Room at the Library of Congress; the staff at the Beinecke Library at Yale; Rebecca Collinsworth at the Los Alamos Historical Society archive; David Fort and the reading room staff at the National Archives; and, a most hearty thank-you to Carolyn Adams and the library services staff of the Mathewson-IGT Knowledge Center at the University of Nevada, Reno—who kept me liberally and efficiently supplied with those teetering stacks of books. If you feel moved to do anything after reading this book, please support your local archives and libraries. If I have learned one thing in doing this project, it is, to paraphrase Henry Adams, that history is just the process of remembering what we have forgotten. That forgetting can have terrible consequences, and the best defense against it I know lies in the archives and in the noble minds and hearts of those who tend them.

Last but not least, I wish to express my gratitude to the University of Nevada, Reno, for the crucial support extended via a Junior Faculty Research Grant from the Office of the Vice President for Research, three Junior Faculty Summer Research Stipends from the College of Liberal Arts, and two Summer Research Assistantships from the English Department.

A note on primary sources consulted for this project: All citations for Francis Bacon's writings, unless otherwise indicated, give volume and page number from the definitive 14-volume edition (1857–1874) edited by James Spedding and colleagues. For interested scholars, the Appendix lists the archives and collections I consulted, along with the type and amount of evidence assembled from those sources.

SCIENTISTS AS PROPHETS

1 PRELUDE

SCIENTISTS AS PROPHETS AND THE RHETORIC OF PROPHECY

"There is no god, and Sam Harris is his prophet." This comment appeared on an Internet forum discussing a new editorial by Harris, a neuroethicist who has published a series of provocative arguments with titles like "The God Fraud" and *The Moral Landscape: How Science Can Determine Moral Values*. Harris claims in these polemics that human well-being is paramount and we should promote moral systems that increase it and dismantle systems that don't. If you haven't read Harris's work or can't guess from the above titles, he believes science increases well-being while religion doesn't, particularly for women and poor people.

While Harris is not presently practicing science, he does hold a PhD in neuroscience from the University of California, Los Angeles. He promotes science-based morality through his nonprofit, Project Reason, as well as on an active lecture circuit. In his 2010 Technology Entertainment Design (TED) conference talk, Harris was by turns witty, tear-jerking, and caustic as he implored his audience to turn to science as humankind's only hope for peace, progress, and fulfillment.

It is not surprising, then, that Harris's performances have triggered responses like the one quoted above. Michael Dowd, a self-identified "evolutionary evangelist" and commentator for Minnesota Public Radio, recently lauded Harris, Richard Dawkins, and Christopher Hitchens as the "New Atheists" who

> are not enemies of religion; they are modern-day prophets. Prophets traditionally were those who chastised their people for having fallen out of sync with their time, with "God's ways." "Come into right relationship with reality," they warned, "or perish!" Today's science-oriented atheists call us

> into right relationship with our time, and that means using all of our best information and cross-cultural experience.[1]

While Harris has not identified himself as a prophet, neither has he rejected any of these recurring attributions. In fact, he calls himself a "heretic,"[2] and he gave Project Reason the evangelistic motto "spreading science and secular thinking."[3] He titled one of his books *A Letter to a Christian Nation*, a phrasing strongly reminiscent of St. Paul's admonitions to the early Christian church.

What I argue in this book about Harris's reception as a prophet, he might find vexing given his sentiments on religion: I argue that Harris's prophetic performances, conscious or otherwise, are far from anomalous; in fact, they're the norm. When Sam Harris steps up to convince the audience at TED or the readers of the *Huffington Post* that science is the only rational basis for human morality, he steps up to an invisible bully pulpit shaped by thousands of years of religious tradition. This stance, which I call the "prophetic ethos," preexisted science as Harris would recognize it by millennia and developed in the inner sancta of the very religions he calls "absurdities." Yet, without the special rhetorical privileges it affords, Harris would be preaching his brand of salvation to the crickets.

The prophetic ethos is a role that a polity—a group of people who must work together to stay together—authorizes to manufacture certainty for them.[4] Now, given that uncertainty is our perennial and inescapable political condition, certainty—in the sense of absolute knowledge—is a chimera.[5] But there is another meaning of certainty that most of us recognize, and that is a sense of conviction. This kind of certainty, which I call "political certainty," is very much achievable, at least for short periods of time. Political certainty is an argument that frames a crisis in terms of "covenant values," which are what I call the values that a polity shares and that distinguish it from its neighboring polities. When we reference American values, scientific values, or conservative values, we are talking about covenant values. It is because we hold these values ourselves (or profess to) and recognize performances of them by those around us that we count ourselves together as a polity of Americans, scientists, or conservatives. These values underpin such arguments as "free markets are best," "the simplest explanation is the best," and "preserving our traditional way of life is best." The temporary political certainty expressed in such arguments can motivate political action or policy.

When a polity encounters a crisis in which right action cannot be ascertained via traditional democratic debate, it turns to its prophets. But while it expects certain knowledge from those prophets, what it gets from them instead is a dialogue that can lead to political certainty. This slippage between the purpose and the result of prophecy will prove crucial to my analysis of our current problems with it in the arena of science advising.

How did scientists come to serve as prophets? Before the seventeenth century, most prophets in the Western tradition were religious intermediaries; they framed dilemmas in terms of covenant values derived from religious authority, thus promoting political action on those grounds. This is the foundation that I examine in chapter 2. But then, through a fascinating series of events and arguments made in Restoration London, among other places, the prophetic ethos was adapted and adopted by natural philosophers, the forerunners of our scientists. I consider this crucial hybridization of scientific and prophetic ethos in chapters 3, 4, and 5.

After that, I track the hybrid of scientific-prophetic ethos through three more twentieth- and twenty-first century cruxes: arguments over nuclear weapons and pesticides in the 1960s (chapters 6 and 7), the popularization of cosmology and evolution in the mass media in the 1980s (chapter 8), and the controversy over climate change in the 2000s (chapter 9). I conclude by suggesting in chapter 10 that if we recognize science advisers as our primary political prophets, this shift in perspective will yield more-practical solutions to the dysfunctions that currently plague our science policy making in areas such as climate change, green energy, and bioethics.

Let me be clear up front about what I am *not* arguing in this book. I am not arguing that all scientists are all priests in the Church of Naturalism, or that scientists *only* perform prophetic ethos when they speak to the public, or that they speak like prophets to each other. Neither would I argue that science advisers are the only prophets we recognize in America. Indeed, one of my primary goals in writing this book is to develop a rhetorical definition of prophetic ethos and to suggest that it can be performed by anyone who can (a) *demonstrate privileged access to knowledge beyond the public ken*[6] and (b) *use that demonstration to engage the polity in a dialogue about its covenant values.* We love our prophets in the United States, and historically they have ranged from environmentalists such as John Muir, to social reformers such as Jane Addams, to scientists such as Rachel Carson, and, naturally, to religious reformers such as Jonathan Edwards. I am treating scientist-prophets in this book because the majority of our political crises today have some scientific or technological aspect, and scientists are our first resort in these crises.

And Now, a Little Greek

I need to explain four terms from classical rhetoric that I employ throughout the book. These are relatively straightforward concepts, and there are no succinct English equivalents for them. The terms are *ethos, kairos, topos,* and *stasis.* I will defer explaining the last two until they are needed (at the end of this chapter and in chapter 5, respectively). But ethos and kairos we need up front.

Ethos (plural: *ethē*) is the root of our word "ethical," and accordingly, it signifies a moral stance, a significant answer to the question "why" about someone's political actions. For Aristotle and his followers, ethos indicated an appeal to the speaker's reputation. As contemporary examples of this ancient usage, think of the endorsements political candidates give each other, or the role Al Gore played in the reception of *An Inconvenient Truth.*

This traditional notion of ethos as personal reputation has been challenged by sociologists and rhetoricians of contemporary public discourse, who have observed that ethos is not a quality emanating from the speaker alone but is rather negotiated between the speaker and her audience over repeated interactions. Robert K. Merton was probably the first sociologist to systematically address scientific ethos in the 1930s and 1940s, and he defined it as a "complex of values and norms" conditioning public behavior and emphasized the social origin of these norms.[7] Rhetorical treatments of ethos have followed in this vein, concentrating on the words and gestures that obey (or violate) the invisible ethical norms.

A recent, influential collection of scholarship, *The Ethos of Rhetoric,* while respecting norm-based definitions of ethos, has recuperated a pre-Aristotelian meaning of the term that connoted "dwelling place" or "abode"—in other words, the rhetorical "place" where an audience would expect to find a familiar speaker.[8] In the ancient tradition, this ethical "abode" supported any of a number of individuals who could perform its hallmarks reliably and recognizably. It is from this revision that I take my working definition of ethos as *a coherent set of expectations about how a person should perform a familiar political role.* These expectations cover both verbal and nonverbal performances, and they frequently reference values—either moral (as in "patriotism") or disciplinary (as in "objectivity"), or both.

The rhetorical situations—debates, speeches, testimonies, comments on blogs, and so on—in which ethos comes into force are often called *kairoi* (singular: *kairos*) by rhetoricians. Kairos stems from a root meaning "climate" in ancient Greek; thus, it carries a meaning similar to our phrase "political climate" but more narrowly focused on events. A kairos is a felt urgency to start a dialogue or the itch to speak up in an opening in an ongoing dialogue.[9] Examples of recurring kairoi in American politics include decisions about whether or not to go to war, trials, victories, reactions to natural disasters, and exultations over the accomplishments of heroes. As should be evident from this list, kairoi often (but not always) open up in response to crises; this is particularly true of kairoi in which a prophet is called to speak up.

Prophets versus Priests, Shamans, and the like

I dedicate the next chapter to explicating the role of the prophet in detail. But a few words of orientation are in order here. I need first to challenge the common notion that the main function of the prophet is to make predictions. Careful examinations of historical and contemporary prophecy have revealed that while prophets are indeed sometimes asked to make predictions, they are just as frequently asked to make decisions about current dilemmas or to uncover truths hidden within the past or present. The unifying factor in all these consultations is not a particular temporal frame but rather the prophet's recalling of the polity to their covenant values.[10] There is no such thing as morally neutral prophecy.

Nor are various prophetic activities easy to pigeonhole. Labels such as priest, sage, and shaman indicate specialized roles in administration, education, and divination, respectively. "Religious intermediaries"—Thomas Overholt's cover term for the people who perform these roles—tend to perform several at a time or in sequence over their careers. I follow Overholt in recognizing the distinction between person and performance, but I use "prophet" instead of "religious intermediary" as my cover term because it is less unwieldy and because I am primarily interested in the actions of *religious intermediaries who are known for their words as much as their works.* I specify what I mean by prophetic ethos in the following discussion; however, readers will rightly recognize shamanic, priestly, or sagacious features in this ethos. For instance, Thomas Lessl recently detailed an ethos performed by modern evolutionists that I would call prophetic but that he calls priestly: the difference in our usage is not categorical but is rather a matter of emphasis, as will become clear in the following chapters.[11] A final note on prophetic terminology: when prophetic offices become cultic (think institutionalized or mainstream), they are generally called oracles, a usage I will observe.

Why Study the Crossings of Scientific and Prophetic Ethos?

If the media so frequently refers to scientists as prophets, and has been doing so for decades, why, then, has it taken this long for scholars to seriously examine this phenomenon? The primary reason is that rhetorical and sociological work on scientific ethos began, quite rightly, inside scientific polities. This literature has documented how scientists negotiate their ethos among themselves, and I address it in chapters 3, 4, and 5.

A smaller group of scholars, however, has studied the tremendous impact of the scientific ethos as a public resource.[12] This work comes from three perspectives. First, political scientists working from a liberal-democratic perspective have voiced concern, at least since the 1930s, about the concentration of political power in the hands of an intellectual elite.[13] Another group of science-studies experts have concluded that in an increasingly complex technological society, reliance on scientists is unavoidable; this work seeks rather to describe and circumscribe, to some degree, this reliance.[14] But a few scholars have argued from a more radically progressive stance that we should break down the barriers between our democracies and our scientific polities.[15] They point to the extraordinary stability, harmony, and productivity of scientific societies over the last 300 years and argue that, to the extent our politicians adhere to the norms of modern scientific ethos, the resulting policies—and thus our society—will be more successful.

In tandem with these secular analyses of the ethos of the science adviser, some historians and sociologists have sought its roots in premodern religious practices. Max Weber developed a theory of social types to explain the development of Western leadership ethos in general. In his account, the *charismatic* priestly leader developed into the *traditional* secular leader, and finally into *rational* authority, as leadership became abstracted from individual personae and attached to political procedures.[16] In Weber's and related accounts, the apparently democratic and rational authority of the science adviser as "opinion leader" in fact derives from the religious and undemocratic authority of the priest.[17] However, derivation does not imply persistence. For these scholars, the Enlightenment marked a watershed, after which truth was attainable through human reason alone without supernatural guidance. Weber famously called this shift the "disenchantment of the world" and treated it as the hallmark of modernity.[18]

While such histories may be true in wide angle, the lines they draw between modern and premodern, scientific and religious, tend to blur when we zoom in to examine particular kairoi in which we call on science advisers. This is the central point of Bruno Latour's short philosophical treatise on the modern condition, tellingly titled *We Have Never Been Modern*.[19] Latour argued from a social constructivist perspective that our modern political crises—over nuclear energy or the origins of life or pesticides or global warming—still serve to goad us into seeking knowledge beyond our ken. When we call on scientists for this knowledge, we place special pressures on them. Thomas Lessl has argued that these pressures result in scientists taking on the characteristics of priests, and in scientific institutions mimicking many of the ethical characteristics of religious institutions.[20]

While I do not argue in this book that *scientists* in general serve as priests of a new natural theology, I do believe that the way we treat *science advisers* is evidence that our political world has never truly been disenchanted. The political dynamics that shaped the bully-pulpit of the premodern prophet, to the extent that they persist, still frame the ethos of the science adviser. I believe that examining the prophetic strains in this ethos can help us see that scholars have significantly mistaken its rhetorical function. If the function of the prophetic ethos is to engage a polity in dialogue that frames a crisis in terms of the polity's covenant values, and if science advisers perform prophetic ethos, then their function is chiefly to start an evaluative dialogue—not, as is often assumed, "pat-answerism," the provision of technical solutions to political problems.[21] Once we acknowledge this misunderstanding, we can explain why the ethos of the science adviser is so unstable—that is, why it predictably oscillates between worship and witch-hunt. We can also predict that proposals to stabilize science advisers' ethos by restricting their policy input to "just the facts" will never work. Further, we can see that the "deficit model" of science communication is bound to fail. This model assumes that the main problem with science policy making is that the "public" doesn't know enough about science; if they knew more, they could better engage in dialogue with science advisers, and policy making would go more smoothly. But if what is really at stake in science advising is not knowledge but values, the deficit model has no traction there. These consequences of my rhetorical genealogy of scientific-prophetic ethos, and others, are considered in more depth in chapter 10.

Rhetorical Genealogy

Now that I have explained what I mean by prophetic ethos and why I believe it is productive to chart its historical crossings with scientific ethos, I should explain how I go about this analysis. In a significant way, a rhetorical genealogy is a genealogy in the common sense of the word—an investigation of ancestry and descent motivated by a desire to explain features or problems we recognize in the present.[22] This kind of genealogy assumes that there are genetic continuities we can pull like a thread through the texture of history, raising interesting bumps or cruxes where key hybridizations with other lines occurred. These crossings are treated as having greater-than-average power in explaining the appearance and behavior of the current generation.

My notion of rhetorical genealogy both shares significant features with and diverges from Foucault's notion of genealogy, which he developed while reading Nietzsche's *Untimely Meditations, Beyond Good and Evil*, and other critiques of historiography.[23] Like Foucault's conception, my genealogy does not seek the origins of the scientific ethos; nor does it assume that this ethos finds its goal or perfection in the present. It does not set reality, identity, or truth as goals but rather looks at how the scientific ethos participates in arguments about these qualities at specific times and places. Like Foucault's notion of genealogy, mine questions the assumed continuity of scientific ethos over the centuries by pointing out moments of significant hybridization with religious ethos, which these days we generally assume to be a different species. I aim to disrupt the perceived purity of the scientific lineage in this way for interested, clinical reasons—to diagnose current problems with the public reception of scientists, such as the men and women who form the United Nations Intergovernmental Panel on Climate Change (IPCC).

Notwithstanding these significant resonances between Foucault's critical project and my rhetorical one, there are some significant differences. In tracing prophetic ethos in public debates about science from the seventeenth century to the present, I assume a degree of continuity that Foucault would reject as a fiction—and not an innocent fiction, but one designed to maintain the privileged status of scientists in our society. But such a rejection is inevitable given my rhetorical focus on ethos versus Foucault's critical focus on subjectivity. A genealogy of ethos assumes continuity. If a polity such as the Royal Society persists, so does the scientific ethos it supports, and vice versa. A genealogy of subjectivity, by contrast, assumes fragmentation at every point, with the subject becoming ever more diverse on inspection, and with the scholar's supposed objectivity falling apart the more closely she inspects her subject.[24]

From Foucault's perspective, a scientific institution like the Royal Society is no more than the list of misidentifications, disagreements, and breakdowns that result each time the name "Royal Society" is articulated. From a rhetorical perspective, that persistent articulation is an act worth investigating *in itself* because repeated rhetorical acts constitute polities, and vice versa. It is true that the Royal Society was never what it said it was; it is also true that it and its audiences kept articulating that identity. It is true that scientists have never been prophets; it is also true that their audiences keep saying they are, and keep working with them toward certainty. I believe the story of this persistence to be worth telling.

Foucault and I would agree that, in any case, a genealogy is not a history. Rather, it is a story, laconic and provocative, about disruptions—in the case of this book, the ways in which prophetic ethos has crossed and disrupted the development of traditional scientific ethos. I believe such a story will shed some light on the problematic situation of science advisers, and I hope it will prompt further inquiries into the phenomenal power of this bully pulpit in our political life. For readers interested in histories of the kairoi I analyze in chapters 2 through 9, the notes will guide you to what I consider to be the best accounts.

The Meme of Prophetic Ethos

Because it is rhetorical and not biological, my genealogy tracks not genes but memes. A meme is a cultural practice that spreads through a population over time via imitation.[25] Although Richard Dawkins developed the term primarily as a way to explain the development and persistence of religion in world cultures,[26] recent research into memes—called memetics—has proven more catholic in scope, treating everything from political slogans to styles of dress to celebrity Tweets. Rhetoricians have focused on the spreading of memes, such as the term "metrosexual," or the use of Western rhetorical figures, such as analogy, in the speech English language learners.[27] Work in this area has constructed a meme as a *persistent, recognizable cluster of rhetorical strategies that is performed in tandem with a recurring kairos and is transmitted via imitation (with mutations acquired in performance)*. To illustrate, one recurring kairos in American life is the required political response to a natural disaster. A recognizable meme commonly performed in such a kairos is the presidential visit to the affected area—complete with speeches, hugs, and the symbolic removal of debris.

A challenge in recent memetic research has been to sufficiently define and distinguish these more complicated memes.[28] Words like "metrosexual" are easy to track with computer searches, but what about the presidential visit or prophetic ethos? In principle, we should be able to identify markers for each meme in the same way that heritable traits can be distinguished by unique signatures of DNA. If we cannot, the meme loses its analytic power, and we cannot track it across kairoi over time.

I hope to contribute just such a memetic method in this project, and I will explain it in detail in the next section. But for readers who want the upshot, I treat the meme of prophetic ethos as a unique cluster of five motivations that distinguish this ethos from closely related ones, such as expert ethos.[29] These motivations are

Ascertainment: The motive to consult a prophet in order to establish certainty in a crisis—its hallmark is evaluation.

Authorization: The motive to establish the prophet outside the polity—its hallmark is polyvocality.

Confirmation: The motive to confirm the prophet's privileged access to knowledge—its hallmark is recognition.

Divination: The motive to descry messages from beyond the political ken—its hallmark is hermeneutics (the conflation of the reading and writing of signs).

Prophecy: The motive to engage the polity in symbolic action that frames the current crisis in terms of the polity's covenant values—its hallmark is dialogue.

In chapter 2, I induce these signature motivations from prophetic practices in the ancient Mediterranean—the Delphic oracle, in particular. To guide the process, I employ a pentadic analysis.

Pentadic Analysis and the Meme of Prophetic Ethos

In *A Grammar of Motives*, Kenneth Burke wrote:

> When things are treated in terms of other things, men may even be said to speak for the dumb objects of nature. Nor are the pronouncements assigned on a purely arbitrary basis. The use of scales, meters, controlled laboratory conditions, and the like, can set up situations in which speechless things can hand down accurate judgments. Men can so arrange it that nature gives clear, though impartial and impersonal, answers to their questions.[30]

Was Burke talking here about the activities of science or prophecy, or both? In fact, he was discussing the dialectic of tragedy, during which the hero's actions are thrown back at him in a way that gives him a transcendent vision of his current dilemma; such a vision, which brings him closer to truth, frequently comes at the cost of his life.

Closely related motivations are at work in the ethē that most concern us in this project—the scientific and the prophetic. In both cases we have self-sacrificing individuals who use technical means to divine transcendent visions of truth. Because of this resonance, and because I have already cast ethos as a "role," Burke's dramatistic pentad provides me with an apt framework for studying the unique dynamics of prophetic ethos.

The pentad just reframes in rhetorical terms the journalistic and scholastic questions that are familiar to almost all of us—who, what, when/where, why, and how. For Burke, humans are "the symbol-using, symbol-making, symbol-misusing animal."[31] All our actions can therefore be seen as symbolic actions, as if they take place on a stage for others to see and interpret. This is the core of Burke's "dramatistic" theory of rhetoric, which answers the five questions above with five key terms:

Who? The *Agent*: Agents are the actors involved in making meaning, including those we would call the speaker/writer and the audience as well as facilitators, critics, and so on. Burke says that nonhuman entities such as ideas may also be agents under the right conditions.
What? The *Act*: This is the symbolic action accomplished. In Burke's model, speech/writing is the prototypical symbolic action, but any action intended or perceived as meaningful counts.
When/Where? The *Scene*: These are elements surrounding the symbolic action that both enable and bound it.
How? *Agency*: These are the means by which the agents act (think instrument, medium, or technology).
Why? The *Purpose*: This is the goal of the symbolic action.

Applying the pentad to the kairos at Delphi, or to any other prophetic situation, is not as simple as identifying the elements that fit the five terms. Burke makes it clear that the real explanatory power of the pentad lies in the dramatic tensions or dialectics driving it. These dialectics resonate between opposing forces represented by antinomies, such as father/son, good/evil, and self/other. Each term in the pentad highlights at least one dialectic, which in turn produces a transformation particular to the drama at hand. For instance, the transformation produced by tragedy, as mentioned above, is a transcendence of the current dilemma which moves the hero on to a new kairos—perhaps death, perhaps a new dilemma. Burke calls transformations such as this one *motivations*—driving forces or rhetorical "engines." There are, of course, many more than five motivations in any given kairos, but to keep our analysis light enough to carry over 2,500 years, the five primary motivations generated by the pentad (listed at the end of the previous section) will sufficiently outline the unique features of the prophetic meme.

Topoi

In applying the five motivations to successive kairoi over time, I am pursuing an ancient rhetorical method, what is often called a "topical method." *Topos* literally means "place" in Greek, as evidenced by our words "topography" and "topology." But classical rhetors elaborated the notion into a system for inventing arguments about sundry subjects; as such, *topoi* became mental "places" a bit more like "viewpoints" or "perspectives." The genius of the system was its codification of regularities in the ways political arguments developed, no matter the subject. For instance, a subject like cutting taxes could be investigated by looking at how tax law had developed over time, how taxes were distributed across the country, the effects of cutting taxes, and so on. You could develop arguments on a different subject—say the freeing of slaves—using these very same topoi of time, space, and cause/effect. In addition to these three common topoi, Aristotle developed a catalogue of 25 more; medieval scholars expanded the list until there were well over a hundred common and special topoi that debaters could use to explore subjects.

These elaborate systems quickly collapsed under their own weight, but the basic concept is still useful—topos as a critical perspective that can keep a scholar's gaze steady and focused across different kairoi over time. Each topos reveals certain patterns in the subject at hand while obscuring others, so inhabiting multiple topoi yields a more complete understanding of a subject—in this case, the question of how the bully pulpit of scientific-prophetic ethos has evolved. During this process, we can expect significant overlap among the topical perspectives, just as Burke predicted for his pentadic terms.[32] The goal of analysis in both cases is a picture of the whole drama, not a neat dissection into five parts.

Reception versus Constitution

What exactly do I use the topoi to look at in a prophetic kairos? Since recent revisions of ethos have emphasized the communal nature of its construction, I might reasonably be expected to focus on the behavior of Agents in each kairos—specifically, what rhetoricians call the audience's reception of a speaker's performance of scientific-prophetic ethos. This was the method I pursued in my first book on scientific hoaxes, and it is the method Leah Ceccarelli used to study the reception of Theodosius Dobzhansky's, Erwin Schrödinger's, and E. O. Wilson's attempts to persuade their scientific colleagues to pursue interdisciplinary projects. Both Ceccarelli and I talked about what our writers "meant" by looking at how their messages were interpreted by readers because, as Ceccarelli pointed out, if a critic only looks at the message itself she "can only say how an audience was *invited* to respond; the critic is unable to make any conclusions about the actual persuasive influence of the text" (my emphasis).[33] It takes a listener for communication to actually occur, and the process of reception unavoidably alters the intended message, as we've all vividly experienced in awkward e-mail exchanges or in childhood games of "telephone."

Although reception studies do valuably expand our view past the writing Agent to include reading Agents, they still cast ethos as primarily conceived and shaped by the writer. Readers have an active role to play in previous reception studies of ethos, but this role is presented as severely constrained by the writer's originary actions and intentions.

Burke's pentad provides quite a different view of the construction of ethos. Ethical dramas result from the interplay not only of writer and readers but also of contextual elements found in the Scene and Purpose and of motivations particular to the Act itself. In other words, if the meaning of E. O. Wilson's arguments for evolutionary biology appeared to us to originate with the writing of his book, Burke would tell us we need to broaden our view: what were the arguments circulating already in Wilson's polity that inspired him to write it? What was the climate in the publishing world that favored its acceptance? What made people want to pick the book up and read it in the first place? It turns out to be very difficult to identify originary moments and motivations for arguments. It makes more sense, then, from a Burkean perspective, to locate the construction of ethos in as many dynamics of kairos as we can lay hands to. I call this broader dramatistic focus *constitution*. In general, I think of constitution as comprising the rhetorical dynamics of social construction.

So, in this course of this study I talk both about reception, the way audiences make meaning from messages they receive, and constitution, how ethos is influenced by all the elements in a kairos, including Scene, Purpose, and modes of Agency. When I consider how particular arguments made by a science adviser were received by a government committee, for instance, I will continue to talk about reception and use primary texts in the way that Ceccarelli and I have in the past. But when I talk about the development of scientific-prophetic ethos across kairoi, I will look to a wider variety of primary and even secondary texts as constitution evidence. These texts will include representations of scientists in the movies or in novels, legislation and legal trials, and opinions expressed by scientists themselves. All these types of texts constrain the ways a particular science adviser's ethos can develop in a particular kairos.

To sum up my procedure then: I first derive the meme of prophetic ethos by applying the pentad to prophetic kairoi associated with the oracle at Delphi and contemporary prophetic institutions in the ancient Mediterranean. Each term of the pentad highlights many dialectics. But for each, I select for analysis a single dialectic that is salient in all the ancient prophetic kairoi in my archive and that in my judgment best typifies that dramatic aspect of prophetic ethos. As the kairoi unfold in contemporary historical accounts, I will look for each dialectic to be transcended by a unique motivation that prophetic ethos provides, one that moves the polity off the horns of its present dilemma and on to a new kairos.

The five transcendent motivations, thus identified and taken together, form the memetic signature of ancient prophetic ethos. Simply put, they specify what a prophet does for the polities that support her. I will then be prepared to look for that special memetic signature in the kairoi that follow in chapters 3 through 9—all of which have been previously identified by scholars as crucial to the development of scientific ethos. At each of these cruxes, I scrutinize primary documents that record the reception of particular performances of prophetic ethos by scientists (natural philosophers, at first); at the same time, I assemble a wider circle of primary and secondary texts that reflect the constitution of scientific-prophetic ethos during the kairos at hand. The five motivations—Ascertainment, Authorization, Confirmation, Divination, and Prophecy—will serve throughout the study as sightlines or threads to pull through the data to track what has persisted, and what has changed, in the meme of scientific-prophetic ethos.

Beginning at Delphi

Even though Foucault has justly warned us that genealogies have no beginning, nevertheless, I have to choose a time and place to start pulling the thread of prophetic ethos, and I have selected the Delphic oracle in its heyday from roughly 800 B.C. to 300 B.C. It was one of the longest-standing and best-respected cultic oracles in the ancient Mediterranean, and during the development of Socratic philosophy, it was amply discussed by historians, playwrights, poets, and philosophers: this rich constitution record is the best archive in which to search for the markers of ancient prophetic ethos.

More importantly, Delphi served Athens, the democracy on which all Western democracies since have been modeled to some degree. Since it's well known that science and democracy have played a crucial and ongoing role in each other's evolution,[34] examining how the oldest democracy in the world related prophecy and policy making will lay a useful foundation for considering how more recent polities have integrated scientists' privileged access to knowledge with policy making.[35]

Having said this, Delphi was of course influenced by other prophetic traditions in the ancient Mediterranean, many of which it operated alongside for hundreds of years. So I reconcile my Delphic analysis with work on prophecy in the region, particularly the Judeo-Christian tradition recorded in the Bible, as biblical prophecy proved a powerful influence on the rhetoric of the early natural philosophers.

In conclusion, here is what I hope to offer in this book that hasn't been offered before, at least not from this perspective and to this depth:

- A rhetorical definition of prophetic ethos[36]
- A method for specifying rhetorical memes and for tracking their ancestry and evolution
- A distinction between reception and constitution studies, and an argument for the latter in studying performances of ethos
- The first serious examination of the popular equation between scientists and prophets
- A redefinition of the ethos of science advisers that has consequences for proposals to address problematic performances of it

In addition, the individual case studies make some contributions of their own that I will point out when we get to them.

But now, enough preamble. Let us turn to Delphi in the golden era of Athens. Let us descend together the dark steps into the *adyton,* the private chamber beneath Apollo's temple, in the company of silent *prophētōn* and the unsteady light of the oil lamps they carry. The Pythia appears suddenly before us seated on her sacred bronze tripod; in her white robes she seems both to glow and to hover in the darkness. There is a sweet smell in the air; maybe it is incense. We feel a little lightheaded. Marshaling our courage, we put to her the question we have come all this way to ask, "What does the prophet do for us?" She shakes the sacred laurel in her hand as the god comes to her, and she responds, cryptically:

Gnōthi seauton. Know thyself.

2 THE DELPHIC ORACLE AND ANCIENT PROPHETIC ETHOS

Not many people know this: in the treason trial that would end his life, Socrates called the god Apollo to his defense. Socrates told the tribunal that his impish friend Chaerephon had years before traveled to Delphi and there petitioned the god's chief prophet, the Pythia: Who is wiser than Socrates? The Pythia's response: No man is wiser than Socrates. Socrates explained:

> When I heard the answer, I said to myself, What can the god mean?... I reflected that if I could only find a man wiser than myself, then I might go to the god with a refutation in my hand... Then I went to one man after another, [and] the result of my mission was just this: I found that the men most in repute were all but the most foolish; and that others less esteemed were really wiser and better... [T]he truth is, O men of Athens, that God only is wise; and by his answer he intends to show that the wisdom of men is worth little or nothing; he is not speaking of Socrates, he is only using my name by way of illustration, as if he said, O men, he is the wisest, who, like Socrates, knows that his wisdom is in truth worth nothing.[1]

With even a passing glance we can see the drama of prophecy in this passage. The pronouncement from Delphi confronted an antinomy—wisdom versus ignorance—and goaded Socrates into seeking the meaning of the prophecy via a symbolic action: questioning his countrymen. This symbolic action yielded Socrates a vision that transcended his original dilemma: "O men, he is the wisest, who, like Socrates, knows that his wisdom is in truth worth nothing." It was a vision that would cost him his life, as Athens's corrupt elite appreciated neither the insinuation that they knew nothing nor the inference that Socrates knew more.

Understanding these nuances of symbolic action at Delphi is the key to understanding the ancient prophetic ethos. The simplest act of the Pythia comprises the entire dramatic pentad—say, her taking a seat on her sacred tripod and drawing a black or white stone from a dish to signify Apollo's yay or nay to a petitioner's query (figure 1). By turning one lens at a time—Purpose, Agent, Agency, Act, or Scene—on symbolic actions such as Socrates's prophecy, I bring into focus five antinomies that lay at the heart of the political dilemmas that sent Athens running to Delphi. Then, I examine how the Pythia's prophetic ethos engaged those antinomies to produce a unique, transcendent motivation that moved Athens off the horns of its dilemma and back into action. Finally, I seek this motivation in accounts of other ancient Mediterranean oracles and prophets. In the end, I will have refined a cluster of five motivations that define the meme of ancient prophetic ethos.

While solving this ancient rhetorical riddle may seem to us now a Herculean labor, while it requires us to delve into the smoky *adyton*, or "forbidden chamber," beneath Apollo's temple and to tease apart the

FIGURE 1 The legendary Attic king Aegeus consults the goddess Themis in her role as the Pythia at Delphi. Note the low ceiling of the *adyton*, the bronze tripod on which Themis is seated, the laurel branch in her hand, and the bowl of lots she holds. The image is from a red-figure kylix in the collection of the Altes Museum in Berlin.

Source: Photo courtesy of Bibi Saint-Pol.

crumbling scrolls of Plutarch and Herodotus, ironically, the answer was engraved for all to see on the temple's portal 2,000 years ago: *gnōthi seauton*. Know thyself, the god challenged all who came to seek his divine wisdom.[2] And how do we come to know ourselves, in the end, but through our actions?

The Oracle at Delphi

As late as 1400 B.C., the airy shoulder of Mount Parnassus, where the oracle would sit, was still serving as a cult site for the earth goddess Ge, whose great dragon Python guarded the sacred Castalian spring that flowed from a cleft in the rocks there (see figure 4a in chapter 6). Delphi was the name Apollo's worshippers gave the site when they co-opted it sometime around the eighth century B.C.: they told the story of Apollo slaying Python and throwing its body into the chasm, where it rotted, releasing the vapors that brought prophetic frenzy to those whom Apollo had taught the art of prophecy. The first of these legendary resident prophets was Themis, the goddess whose name meant "custom," "tradition," and "divine law." Then Apollo installed a human priestess, the Pythia, named after the guardian python he had slain.

Between 800 B.C. and 400 B.C., Apollo's temple was repeatedly refurbished and enlarged. Eventually it was hedged in by a maze of statuary, shrines, treasuries, and even an amphitheater that hosted regular athletic and dramatic performances; in short, Delphi became a Hellenic "entertainment complex."[3] But the political and religious winds inevitably shifted, and despite intermittent attempts by enthusiastic Roman emperors to restore the oracle's preeminence, it steadily declined until Theodosis I closed it permanently in A.D. 395.

The star performer in the drama at Delphi, the Pythia, was to the best of our knowledge a middle-aged woman selected from a pious peasant family in the region.[4] She held forth in the womblike *adyton*, seated on a bronze tripod above the chasm that had swallowed Python. Near at hand hulked the *omphalos*, the "navel" stone that Greeks believed marked the center of the world. The Pythia was attended by male *prophētōn*, after whom our own prophets are named. The word *prophētēs* in Greek meant "speaking forth," "speaking for," or "speaking before"; as part of their duties, the *prophētōn* recorded the Pythia's *chrēsmoi*, or pronouncements, for petitioners to take home with them.[5] In the following sections, I examine a series of these recorded consultations to identify how the Pythia's prophetic ethos impacted the drama in each case.

Ascertainment: The Lens of Purpose on the Motivations of Prophetic Ethos

Sometime shortly before 546 B.C., King Croesus of Lydia became anxious about Cyrus's encroaching Persian armies. According to Herodotus, the king decided that since he would soon need advice from the gods about what to do, he should check all his divine channels to make sure they were open and functioning. So, he sent messengers out from Sardis with the order that exactly 100 days to the minute from the time of their departure, they were to ask all the major Greek oracles (and the oracle of Ammon in Libya) what King Croesus was doing at that moment. Herodotus writes of this televisionary petition:

> Now none relate what answer was given by the rest of the oracles. But at Delphi, no sooner had the Lydians entered the hall to inquire of the god and asked the question with which they were entrusted, than the Pythian priestess uttered the following hexameter verses:
>
> I know the number of the grains of sand and the extent of the sea,
> And understand the mute and hear the voiceless.
> The smell has come to my senses of a strong-shelled tortoise
> Boiling in a cauldron together with a lamb's flesh,
> Under which is bronze and over which is bronze.[6]

And indeed, at the appointed time, Croesus had cooked a tortoise and a lamb together in a lidded bronze kettle; he praised Delphi as the only true oracle. Unfortunately for Croesus, this happy result immediately led to a disastrous consultation: Croesus asked the oracle who would win if he faced off with Cyrus and was told that if he crossed the River Halys, a mighty empire would be brought down. Assuming that the Pythia meant Cyrus, Croesus attacked and was routed. Stunned, the Lydian king sent to Delphi to demand of Apollo "whether he was not ashamed to have encouraged Croesus by his oracles to invade Persia in the confident hope of destroying Cyrus":

> The Pythia replied that Croesus had no reason to complain about the oracle: the god had declared that if he attacked the Persians, he would bring down a mighty empire. After an answer like that, the wise thing would have been to send again to enquire which empire was meant, Cyrus's or his own. But as he misinterpreted what was said and made no second enquiry, he must admit the fault to have been his own. (1.90)

If we apply the lens of Purpose to this pair of consultations—if we look for the motivations for beginning the drama—we squarely confront the classical antinomy of actuality versus potentiality. Actuality described how things were; potentiality described all the possible ways they could be but weren't. Aristotle used the terms to distinguish eternal elements of the universe like the sun from changeable beings like humans.[7] This antinomy persisted into the kairoi we will consider in the seventeenth century, creating grave political problems for astronomers like Galileo and Copernicus who observed changes in the supposedly unchanging heavens. And it created problems for Croesus as well. In his tests of the oracle, the king thought he had found a clear window onto his future; but in the disaster at Halys, he learned that his attempts to fix the future state of things *ironically changed them*. The "big-C" Certainty that Croesus thought he had gotten from Delphi in each case was his ostensible Purpose for consulting the oracle. But the Pythia intervened in this original Purpose and led Croesus in a different direction—toward a "little-c" certainty that framed his political dilemma in terms of his core values. Croesus's Ionian superiority complex led him to underestimate the "barbaric" Persians, and his generosity to Delphi convinced him he could rely on Apollo's assistance in battle. Through his defeat and the Pythia's gloss on it, Croesus was brought to a transcendent and tragic vision of his arrogant certainty.

To sum up, the Pythia transcended the actuality/potentiality antinomy by showing petitioners that their assumptions and actions *made* the future and thus made it unknowable. She took their attempts to grasp Certainty and rechanneled these into political actions that made petitioners acknowledge the values they lived by. I call this special motivation the Pythia provided Ascertainment. Cicero recognized this motivation in Roman prophetic practices as well. Although he presented himself as sharply skeptical of miracles in "On Divination," he had himself functioned for a time as an official augur for the Senate. Augurs were responsible for reading the feeding behavior of sacred chickens or the patterns of lightning strikes for omens favorable to elections and other legislative functions. Cicero pointed out that such attempts to know the future ironically and surely changed it by changing the behavior of its participants. Nevertheless, he acknowledged the political "expediency" of augury in settling protocol dilemmas and instilling in the Senators a provisional sense of certainty that moved them to take action.[8]

This exchange of certain knowledge for conviction has been observed in the practices of other ancient Mediterranean oracles and beyond. A few examples: ancient Judeans asked Jeremiah for certain knowledge of an impending Babylonian invasion and instead were called to repent and return to Yahweh;[9] ancient Egyptians followed the swaying of a paraded effigy of Anubis for answers to their murmured questions and saw in those motions the answers that already lay in their hearts;[10] ancient Mari court seers tended to give predictions that validated policies already in place;[11] and down to the twentieth century, cargo cults have sought to learn the future from Western technologies and objects and, by so valuing these things, have ensured that their future will be an increasingly Western one.[12]

These examples illustrate how the ancient prophetic ethos settled dilemmas. It could not provide the kind of certainty the polity wanted it to (absolute knowledge); nonetheless, it helped the polity manufacture a provisional political certainty that could move it off the horns of its dilemma and into a new kairos.

Authorization: The Lens of the Agent on the Motivations of Prophetic Ethos

Sometime around the year 508 B.C., as part of the reforms that ushered in the style of democracy we have come to associate with Athens, Cleisthenes petitioned the Pythia to select from a list of 100 founding heroes the 10 that would serve as mascots for the new military regime he wanted to establish. The Pythia selected for this honor Hippothon, Antiochus, Ajax, Leo, Erechtheus, Aegeus, Oeneus, Acamas, Cecrops, and Pandion. Some were historic heroes and some legendary; many were generals themselves. All had demonstrated an undeniable zeal for carving out a Greek nation state from surrounding people groups, often violently.

The kairos that contained this petition was the reformation of Athens. If we look at who the Agents are in this drama of political self-definition, the primary antinomy we encounter is self/other (*autos/allos*). The Greeks invented logic, a philosophical system one of whose fundamentals is the recognition of difference. Self/other was probably the first of these divisions, marking off subjectivity from objectivity, human from human.[13] Once this difference could be expressed in language and logic, it could be politicized. And from there, it was a short step to arguing that difference gave "us" rights that "they" were not entitled to, a step the West has since marched to over and over in its various imperialist projects.

But critical awareness of this causal chain leading from the self/other antinomy to colonization did not have to wait for the postmoderns: the sophist Isocrates expressed exactly this sentiment in "Hymn to the Logos," in which he claimed that the *logos*—which can be translated as "logic," "speech," "ratio," or "measurement"—had literally built the polity by binding Athenians together linguistically, culturally, and politically.[14] He further contended in the

Antidosis that this accomplishment justified Athens in extending its dominion over the Other of the woman, the slave, the foreign.[15]

This antinomy puts an ironic frame around the authority that the Pythia wielded in the drama of Cleisthenes's petition. First, the Pythia herself was *xena*, a non-Athenian "stranger," and, secondly, a woman—so, she represented the very groups who had been traumatized by nearly every name on Cleisthenes's list. To add insult to injury, the majority of Athens's early consultations with the women who held her office (750–500 B.C.), and the majority of Delphic political consultations overall, sought Apollo's blessing for acts of colonization or war.[16]

So, Delphi was decidedly Other to Athens.[17] But the Pythia's situation was hardly unique: almost all prophets in Greece were foreign to the polities that consulted with them. And many interstate cults functioned under the auspices of powerful women.[18] In a fragment from Euripides's lost play *Melanippe*, one of the female characters offered women's traditional association with the divine as evidence of their political power: "Now as for our dealings with the gods, which I judge to be of prime importance: we have a very great share in them. In Phoebus's halls it is women who proclaim the mind of Loxias, and around Dodona's holy foundations by the sacred oak it is the female sex that purveys the thoughts of Zeus to any Greek who seeks them."[19]

How do we reconcile this flattering contemporary portrait with the Pythia's participation in Cleisthenes's glorification of those who had brutally colonized her people? Marie Mauxion attempted to explain this conundrum by analyzing the language of the characters of Apollo and the Pythia in Euripides's plays.[20] Mauxion found that Apollo's warlike, confrontational discourse differed enough from the Pythia's verbs of nurture to suggest that Greek audiences recognized (and feared) the Pythia's differing discourse of the earth and the feminine. In other words, her voice was recognized even though it could not be extricated from Apollo's or from the voices of the male *prophētōn* who recorded it.[21] What Mauxion's work reveals is that the Delphic oracle was irreducibly polyvocal; it spoke at the same time for Athens and for the Others that Athens struggled to dominate and define itself against.[22]

This observation can help us understand how the Pythia's prophetic ethos intervened in Cleisthenes's petition. The debate over which heroes best modeled the covenant values of Athens could have gone on forever. But the Pythia's undeniable connection to the Other (female, foreign, supernatural) provided the kind of external authorization that Athens needed when its internal negotiations had ground to an impasse. She picked 10 names from the list of 100, and Athens could move ahead with organizing its democracy. Her response transcended the self/other antinomy by appearing to shift the grounds for judging between the two from the political to the extrapolitical, from the natural to the supernatural. Yet at the same time, she was voicing Athenians' own covenant values back to them—the cluster of mores, fears, and desires that made them unique as a polity.[23] Athenians heard themselves recalled through her foreign tones to embrace these values anew. The resulting sense of conviction catalyzed political action—war, colonization, the building of another temple, or the offering of a sacrifice. By these actions Athenians made themselves known to the Others around and among them. At the same time, they were reminded of the presence of these Others: the woman crying out from the *adyton*, the slaves carrying the sacrificial piglets around the Assembly, the foreign *chrēsmōdoi* (oracle singers) chanting prophecies at the command of the Council. Thus, citizens recalled the values that defined them as a polity: we are the people who honor Apollo; we are the people who fear the power of the Other; we are Athens.

I call this motivation that the prophetic ethos contributed to ancient dramas of self-definition Authorization. It is a motivation visible well beyond the bounds of Delphi. The Pythia was a cultic prophet, associated with a state religion and shrine, but there were many wandering prophets, whom the Greeks called *manteis*, a word that Plato said derived from the *mania* that seized them as sensible evidence of their coming into the presence of the divine.[24] Socrates's legendary teacher Diotima was perhaps the perfect instantiation of this polyvocality. In the *Symposium* she was called a "Mantinean" priestess—a fabrication that ingeniously and simultaneously inscribed her foreign origins and her *mantike*, or prophetic, power. Socrates claimed she taught him the "love arts" and the art of persuading publics.[25] His implicit argument that political discourse had religious roots was made explicit by the Greek historian Strabo, who argued that the first city-states on the Attic peninsula had coalesced around oracles and temples:

> Such being the advantages of the site of Delphi, the people easily came together there, and especially those who lived near it. And indeed the Amphictyonic League was organized from the latter, both to deliberate concerning common affairs and to keep the superintendence of the temple more in common, because much money and many votive offerings were deposited there, requiring great vigilance and holiness.[26]

Here we can again see the way in which prophetic Authorization transcended the self/other antinomy by entangling the first Greek polities with the female and foreign Other at Delphi in the drama of self-definition.

We can see a similarly transcendent motivation at work in prophetic dramas across the ancient Mediterranean—particularly in the way they developed over time. Most nomadic tribes first consulted with wandering prophets like the *manteis,* who appeared in times of turmoil and then moved on when the kairos resolved; I call these prophets "kairotic prophets." Then, as nomadic tribes settled and coalesced into agrarian polities, prophets either settled down with them to form the polity's dominant cult or were pushed to the margins, from which they continued to voice the concerns of the poor, the slaves, the disabled, and the colonized who occupied that political space.[27]

Even the cultic prophetic offices continued to mark their Otherness both physically and socially. Ancient Israelite and Amorite oracle tents were erected outside the main campsite.[28] Israel's cultic prophets frequently went into the wilderness to seek communion with God or to escape persecution from state authorities. Jonah, the story goes, worked for a while from remote offices, including the belly of a whale and the shade of a castor-oil plant (1:17; 4:5). Socially speaking, though some ancient Mediterranean prophets were scions of priestly or noble families, most came from humble origins.[29] Women were much better represented in these offices than they were in other political offices across the ancient world.[30] So, even what appeared to be the ideological centers of many of these ancient polities had been reverse-colonized by Others.

To sum up, in Greece and across the ancient Mediterranean, Agents in prophetic interactions were fundamentally ambiguous. They were foreigners but were treated as authorities. They stood outside the polity but sustained and were sustained by it. They voiced in foreign accents the polity's very own fears and values. These are the paradoxes that prophetic ethos transcended via Authorization.

Confirmation: The Lens of the Scene on the Motivations of Prophetic Ethos

In "The Obsolescence of Oracles," Plutarch related a harrowing episode that occurred at Delphi not long before he became a *prophētēs* there. A foreign dignitary had come to consult the Pythia, but the sacrificial goat would not nod its head in assent and had to be "nearly drowned" before it gave even a hint of the necessary sign. Despite this bad omen, the *prophētōn* cajoled the Pythia to take her place in the *adyton*:

> What, then, was the result touching the priestess? She went down into the oracle unwillingly, they say, and half-heartedly; and at her first responses it was at once plain from the harshness of her voice that

> she was not responding properly; she was like a laboring ship and was filled with a mighty and baleful spirit. Finally she became hysterical and with a frightful shriek rushed towards the exit and threw herself down, with the result that not only the members of the deputation fled, but also the oracle-interpreter Nicander and those holy men that were present. However, after a little, they went in and took her up, still conscious; and she lived on for a few days.[31]

When we look more carefully at the scene—the written and unwritten rules that constrain the drama—we immediately see questions of order and the breaking of order, of making sense and losing it. These questions belong to an Ancient Greek debate over the dividing line between the sensible and rational world on the one hand, *kosmos,* and what was unknown and chaotic on the other hand, *xaos.* The word *xaos* carried the meaning of today's "chaos" as well as the smoke of an older root meaning "that which burns." But separating cosmos from chaos at Delphi was no easy task. After all, Prometheus, the Titan who gifted the Greeks with prophecy—the very skill that allowed them to put in order the chaotic swirl of nature and fate around them—also gave the gift of fire. Fire enabled both the forging and the destroying of civilization, as Heraclitus was fond of reminding his fellow Greeks.[32] So, just as with the self/other antinomy, the cosmos/chaos antinomy inscribed the terror of annihilation.

Thus, the Greeks who consulted Delphi watched the Pythia's behavior closely through the laurel smoke for strangenesses—strains of chaotic illegibility that demonstrated that Apollo had overwhelmed her rational mind and was speaking through her. This state was called *enthousiasmos,* (god filled), the word from which we get "enthusiasm." In deciding what counted as rationality and what counted as madness in the Pythia's performance, petitioners were trying to mark the boundaries between cosmos and chaos.[33]

But if we look again at Plutarch's account of the frenzy and death of the Pythia, we see that it provided a unique motivation that transcended the Greeks' obsession with separating cosmos from chaos. For Delphi to function successfully as an oracle, petitioners had to bring more with them than money and goats: they had to bring a willingness to recognize chaotic events as ordered and significant. Movements of the goat's head must be interpreted as "nodding"; changes in the baseline of the Pythia's tone must be remarked and described as "harsh" and "hysterical"; her speedy exit from the *adyton* must be linked causally to the breaking of the rules of consultation; so must her death. All these moments must be recognized as Apollo's symbolic actions and a message inferred: I will not be ordered by humans.

We cannot overlook the power of the scene in Apollo's temple to focus and shape the drama of recognition. The petitioners have come here only after a series of ritual sacrifices, painful expenditures, and a frustrating wait, all of which have heightened their anticipation and made them hypersensitive. The *adyton* is dark and low, so they have to stoop. The sole furnishings here are the tripod and the *omphalos* covered with an arcane white net of wool. When the Pythia enters, she is dressed in vestal white, trailing the acrid smell of purifying laurel smoke; there may also be the ethereal scent of ethylene; robed and silent *prophētōn* attend her.[34] The scene is supercharged with recognition, pregnant with the hybridization of cosmos and chaos. Anything the Pythia says or does in it is likely to be interpreted as supernaturally motivated.

The prophetic function we see in the Scene at Delphi is Confirmation. The Pythia engaged petitioners in a drama of recognition that mooted the cosmos/chaos antinomy. The petitioners' willingness to attend to chaotic events as cosmologically relevant predetermined their experience in the *adyton*. They had come to seek Apollo's answers to their questions; so doing, they had again and already confirmed the prophetic ethos of the Pythia. And so we now see in a new light the ambiguity of the name of Themis, Delphi's first oracle, which meant both divine law and human tradition. By recognizing divine will and taking political action in accordance with it, the Athenians crossed chaos and cosmos in ways that could not be untangled.

Thucydides dramatized the effects of such crossings in his account of the Athenians' debate with the Melians during the Peloponnesian War. The Athenians claimed that they were driven to induct surrounding peoples into the Athenian cosmos by a divine directive: "Of the gods we believe, and of men we know, that by a necessary law of their nature they rule wherever they can. And it is not as if we were the first to make this law, or to act upon it when made: we found it existing before us, and shall leave it to exist for ever after us; all we do is to make use of it, knowing that you and everybody else, having the same power as we have, would do the same as we do."[35] The word for "law making" here in the Greek is *chrēsamenoi*, from the same root as *chrēsmoi*, the Pythia's pronouncements. By using it, the Athenian delegates perform the inextricability of cosmos and chaos, of law and fate. Ordering chaos is fundamentally an act of cosmology, and cosmology is a political action confirming the authority of the current lawmakers. Thus, Confirmation represents a circular function by which lawmakers are authorized to make cosmos from chaos; meanwhile, the ability to make cosmos from chaos is taken as a divine sign of law makers' right to rule.

Scholarship on the wider setting of prophecy in the ancient Mediterranean corroborates the fundamental circularity of Confirmation. Thomas Overholt, who first used the term to describe the politics of recognition in prophecy, pointed out its central irony: the utterances of ancient prophets had to meet their audience's basic formal expectations for what these communications would be like; if the prophets' pronouncements were not recognizable to their audiences at least in part, these utterances would simply be lost in the background noise of chaos.[36]

If we survey the literature on Confirmation, we see that certain types of symbolic actions were commonly interpreted as signs that a polity had correctly authorized its prophet, that the prophet was correctly "hooked up," so to speak, and was receiving authentic messages from the divine. These symbolic actions performed by the prophet included the following: a story about a divine "calling," an ascetic or marginal lifestyle, compulsory mantic behavior (acts indicating possession or ecstasy, such as the Pythia's "altered tones"), the uttering of prophecies, allegorical performances, and miracles.

First and most importantly, the evidence of a "calling" functioned to select the prophet from his/her community. These "callings" often occurred during adolescence, as they still do today in polities that support prophets.[37] In ancient Israel, a prototypical example is the young Samuel waking to hear God calling his name, and then going to the priest Eli to confirm the call (1 Sam. 3:1–9). By contrast, ancient Greek *manteis* were often selected more actively by the community; in some cases, a more experienced *mantis* would spit in his disciple's mouth to transfer the prophetic gift.[38]

Next, an ascetic lifestyle marked the prophet as standing outside politics as usual and in a special relationship with the divine. The hardships of that lifestyle were judged as divinely inspired; why would any normal member of the polity choose to put on such a difficult performance, year after year, by his/her own volition? Asceticism is ubiquitous across ancient traditions: two quick examples are John the Baptist's diet of locusts and honey (Matt. 3:4) and the *selloi* of Zeus's oracle at Dodona, who slept in the dirt the better to receive the god's communications from the earth.[39]

The ascetic lifestyle properly belongs to the larger category of allegorical performances, which the prophet often followed with an interpretation that fit them to the present crisis. Some stand-out examples from the Hebrew Bible include the prophet Hosea's taking a prostitute as a wife and having three children by her (Hos. 1:1–11), and Ezekiel lying on his left side for 390 days and his right for 40 days (Ez. 4:4–8). Both of these performances signaled details of God's judgments against the polities of Israel and Judah.

Compulsion was a key feature of these and other confirming signs of prophetic ethos. Hosea and Ezekiel stressed that God compelled them to their bizarre performances. Compulsion was often a key feature of the "calling" narrative as well: if the candidate resisted the call, the deity would punish her with madness.[40] In general, any manic performance, including the Pythia's putative frenzies, was taken as a sign of divine compulsion.

Prophecies will be discussed in detail below, but they functioned as confirming signs in both content and form. Prophecies were always marked by atypical speech patterns—whether these were poetic elements, signal phrases, or, less commonly, frenzied delivery. These elements were perceived as being beyond the prophet's individual ability and thus confirmed the divine source of the utterance.[41]

Miracles are the flagship confirming actions of prophecy. Some show-stopping prophetic miracles from the Hebrew Bible include Moses visiting Pharaoh with 10 plagues (locusts, boils, frogs, etc.) to persuade him to release the Israelites (Ex. 7–11), and Elijah calling down fire from heaven from atop Mt. Carmel (1 Kings 18). But miracles were attested as prophetic even in cosmopolitan Rome. In Cicero's dialogue *On Divination*, his interlocutor, Quintus, assembled an impressive list of omens including Jove's striking the Capitol with lightning, a razor cutting a whetstone in two, and a Spartan statue of Hercules dripping with sweat just before the ill-fated battle of Leuctra (2.12, 17, 33).

To sum up the special prophetic motivation of Confirmation at Delphi and elsewhere in the ancient Mediterranean, it was largely a politics of recognition.[42] Audiences would not recognize an utterance, allegorical performance, or miracle that deviated significantly from their expectations regarding the form and content of these symbolic actions; any such deviations were attributed to madness, trickery, or corruption.[43] At the same time, Confirming performances did exactly that—cemented the prophet's special authority to shape the polity's cosmology. These circular cosmological operations were developed and internalized by polity and prophet during the prophet's Authorization: in other words, we need to remember that the prophet's selection, location, and support dictated to a great extent how she would and could communicate with the polity. Therefore, Authorization and Confirmation are reciprocal functions in the social context of prophetic ethos.

Divination: The Lens of Agency on the Motivations of Prophetic Ethos

In perhaps its most technically spectacular Delphic consultation, Athens petitioned the Pythia to determine if it should put under cultivation borderlands called the Sacred Orgas near the Mysteries at Eleusis. These lands had traditionally been left wild as a memorial to the two goddesses of the

Mysteries, Demeter and Kore, who had taught the Athenians agriculture. These peaceful-sounding borderlands were nonetheless a constant bone of contention between Athens and Megara, some of whose farmers had squatted there. Around 352 B.C., Athens and Megara were once again on the brink of war. Athens proposed to consult Delphi as a way of easing these tensions, but either Megara or the Athenian Assembly must have voiced some qualms about the typical procedure, because the Athenian delegation took some truly extraordinary precautions with the consultation that followed.

The two options to be presented to Apollo—leasing the borderlands to pay for restorations to the Mysteries or leaving the lands uncultivated for the goddesses—were written on tin tablets, which were wrapped in pure wool and then shaken inside a bronze jug. The leader of the Assembly put in his hand and took out each concealed option, placing one in a gold jar and one in a silver jar. The jars were sealed with the public seal of Athens, and then the Assembly leader invited any citizen who wished to add his seal. The jars were carried to Delphi and presented to the Pythia, and the jar she chose under Apollo's inspiration was returned to Athens and opened in the Assembly with a flourish. The god had ordered the lands to remain uncultivated.

The Sacred Orgas consultation may have been unusually elaborate, but its primary mechanism was the default mode for Delphic consultations—presenting the Pythia with two options for Apollo to select between. Given my argument about how important antinomies were to the ancient Greeks, this practice should not come as a shock.

Of the over 600 extant Delphic oracles that historian Joseph Fontenrose collected from scrolls, temple inscriptions, and historical accounts, the majority present an idiomatic form: *lōion kai ameinōn*, "better and more advisable." Using this formula, the petitioner would present to the Pythia either one or two plans.[44] If two, the plans were sometimes engraved on bone or wooden lots for her to choose between, sometimes simply offered as a verbal pair; rarely, they were presented as tamper-proof artifacts, as in the Sacred Orgas consultation.[45] To the plans presented by this formula, the Pythia would indicate Apollo's choice by echoing the *lōion kai ameinōn* formula, or she would select a black or white token (a stone or bean) from a bowl she held in her hand, as depicted in figure 1; this was also the custom at Zeus's oracle at Dodona. The *prōphetōn* would inscribe the god's response for the petitioner to take with him, once more repeating the formula: for example, "It is better and more advisable that you sacrifice to Demeter."

When we apply the lens of Agency to the Sacred Orgas consultation and others like it—when we ask, How is the drama enacted?—what snaps into focus is the classical antinomy between reading and writing, the reception and the production of signs. Sophists and Socratic philosophers in the fourth century B.C. drew a distinction between writing and speaking (*graphein* and *legein*) on the one hand and reading and hearing (*anagignoskein* and *akouein*) on the other.[46] And the advice they gave students tended to shift the power in this drama firmly toward the speaker, who was active and productive while his audience was passive and receptive.[47]

The Sacred Orgas consultation, and all the *lōion kai amenon* consultations at Delphi, mounted a serious challenge to these contemporary attempts to privilege the production of texts and segregate it from reception. Athens was not just a passive *reader* of the god's will but a coauthor in *writing* it, since they preselected the options the Pythia chose between. Even when we back away from the rigidity of the *lōion kai ameinōn* formula, we find no consultation in Fontenrose's collection that is not framed by a question of some sort from the petitioner. Consultations with the Pythia thus transcended the reading/writing antinomy by implicating petitioners in writing their own realities and futures—just how dramatically, we will see shortly, when we consider the Pythia's amphiboles.

This special motivation that prophetic ethos provides I call Divination, after Thomas Overholt's term for the instrumentality of prophecy.[48] Divination scuffs out the line that polities try to draw between the production and reception of signs or between the Agents of these different acts. In the Greek tradition, this dynamic was visible even in origin myths about Divination. In Greek myth, Prometheus and Apollo taught the craft of Divination to humans. The type most closely associated with Prometheus was *hepatoskopia*, or reading the creases in sheep's livers. Ancient sources indicate that entire flocks of sheep were driven alongside Greek armies so their attendant *manteis* could use them to divine auspicious military strategies. Herodotus related an occasion during the battle of Plataea in which the Spartan army was held up for days under heavy fire from Persian archers while waiting for the livers to sanction an advance (9.61–62). Greeks also employed *ornithomanteis* and *exēgētōn* to read the flight of birds and other omens. *Oneirokriteis* judged dreams, particularly those had just before waking, as portentous. But there is nothing about the flight of birds or the shape of livers that necessarily looks like Greek letters: "seeing" these signs required creative recognition.

This conflating of reading and writing is nowhere more apparent than in cledonomancy. In the origin myths about *klēdōnomanteia*, or the interpretation of chance utterances, Hermes, the "messenger of the gods," played a key role. When Apollo refused to teach him Divination, the story went, Hermes complained to Zeus, who allowed his messenger the ability to divine fate from seemingly random motions.[49] So, in the most common practice of cledonomancy, a "Hermes pillar" topped with the god's head was erected in the town marketplace. On each side of the pillar were inscribed numbered prophecies. Petitioners would roll special four-sided dice made from sheeps' knuckles in order to determine which prophecy pertained to their situation. Sometimes, these prophecies gave further instructions about which other divinities to consult for help, such as "pray to Athena." But sometimes they were fortune-cookie-like statements such as "look on your left." This may seem like the Magic Eight Ball of the ancient Mediterranean, but it was taken very seriously. St. Augustine reported that his conversion to Christianity was prompted by a cledonomantic instance; sitting pondering the fate of his spirit in a friend's garden, a copy of the gospels lying near to hand, he heard a little boy on the other side of the garden wall sing, "Pick it up and read it."[50] For Augustine, as for any interpreter, then came the challenge of interpreting the randomly encountered prophecy in light of his current dilemma. Enter *hermēneutikē*.

With the proliferation of oracles and *manteis* in Greece, a service industry sprang up: expert *chrēsmologoi* archived prophecies for the Assembly in Athens and *chrēsmōdoi* were hired to recite them on demand. For instance, around 340 B.C., Demosthenes wrote a speech for a client, Sositheus, in which Sositheus was to call in the Assembly for a *chrēsmōdos* to recite a Delphic prophecy as evidence against his opponent Macartatus.[51] The role of professionals in the interpretation of texts generally came to be called *hermēneutikē*, from which we get our word "hermeneutics"; the shadow of the Hermes pillar still falls across the term.

Heidegger revived this prophetic sense when he explicated his own hermeneutics over two millennia later.[52] Hermes's mischievous spirit, his liminal status between the human and divine worlds, and the boundaries that cledonomancy blurs between the writing and reading of signs—these all infected Heidegger's conception of the role of the literary critic. His student Hans Georg Gadamer, extending this line of inquiry, insisted that any critical reading or "understanding" of a text was actually a writing; therefore, production/reception was a false antinomy in literary criticism.[53]

I would argue with Gadamer that the key to understanding ancient Divination lies in understanding that it resided in the space between reading and writing. When the gods spoke, they did so through the medium the diviner had chosen, whether that medium was words, beans, a liver, bone lots, or elaborately sealed jars. In fact, research on divinatory practices across the ancient Mediterranean confirms the Delphic ambiguity of hermeneutics. The book of Deuteronomy includes a laundry list of regional divinatory specialists whom the Israelites were barred from consulting—despite the fact that their own priests employed many of the same techniques in their official functions. These techniques included casting lots and divining by the sacred jeweled breastplate, or *ephod;*[54] all relied on the priests inscribing chaotic motions, sounds, or other changes as legible messages from Yahweh. Similarly, Babylonian sages read the motions of the planets as spelling out the future of their king, and Syrian court seers cast bone dice to select inscriptions carved on temple walls in the same way that petitioners used Hermes pillars in Greece.

To sum up, Divination is the special motivation provided by prophetic ethos to transcend the reading/writing dilemma through hermeneutics. Through Divination, the petitioner and god enter into a dialogue that creates political reality now and in the future.

In the *Symposium*, Plato reported a speech his Mantinean teacher Diotima once made to him about love. She claimed love was a *daimon*, a Hermes-like messenger who belonged to the human and divine worlds equally: "interpreting and transporting human things to the gods and divine things to men; entreaties and sacrifices from below, and ordinances and requitals from above: being midway between, it makes each to supplement the other, so that the whole is combined in one" (202E). In just this way are Divination (human questions to the divine) and Prophecy (divine words to humans) inseparable. Consequently, as I have been considering the hermeneutics of Delphi, I have been talking about Prophecy at Delphi as well.

Prophecy: The Lens of the Act on the Motivations of Prophetic Ethos

Now it is time to consider the Pythia's most famous prophecies: amphiboles and riddles. *Amphibolos* means "double-pointed" and indicates an ambiguous response such as Croesus's River Halys prophecy. The Pythia's riddles were poetic puzzles just like the ones we pose to each other for entertainment today. While these forms accounted for only a tiny fraction of the Pythia's recorded oracles, they featured disproportionately in contemporary accounts because of the dramas they instigated. Historians disagree about whether the Pythia actually spoke them in such elegant form—or indeed, if they were uttered at all; however, as a rhetorician, I am more interested in *why* these apocryphal prophecies fascinated contemporary commentators. I believe they persisted in historical accounts because they reminded the Greeks (and us) of the impossibility of separating belief from action in the process of self-definition.

In Book 3 of his *Rhetoric*, Aristotle discouraged rhetors from indulging in amphiboles unless they were actively trying to make their argument obscure.

He ribbed the seer Empedocles for "pretending to say something" through amphiboles, thus increasing the odds that his predictions would align themselves with the actual outcome of events (3.5.4). We see similar types of ambiguous statements in newspaper horoscopes today.

There are perhaps a dozen of these types of responses in Fontenrose's catalog of Delphic pronouncements. In one, King Aegeus asks the Pythia if there will be an heir to his throne, and she responds, "Do not the wineskin's salient foot untie...until you come to hearth and home again"; the "salient foot" turned out to be a double entendre, which Aegeus discovered only after he bedded Aethra on his way home to Athens and sired the bastard hero Theseus.[55]

King Croesus of Lydia seemed to collect amphiboles—the sharp ends of them, at any rate: the most famous, which we have already encountered, was the prediction that if Croesus crossed the River Halys, "a great empire would be brought down." That empire turned out to be his own (Aristotle, *Rhetoric* 3.4). Another Croesian amphibole predicted that his kingdom would stand until "the Medes have a mule as a king" (Herodotus 1.55); judging this eventuality impossible, Croesus attacked Persia and was soundly defeated, as Cyrus was half Mede, half Persian.

Aristotle discusses another Delphic amphibole in the *Rhetoric* in a different light—not as a comeuppance for arrogant kings, but as a goad to political dialogue. "For example," Aristotle wrote, "Themistocles [interpreted] the 'wooden wall' to mean that a naval battle must be fought" (1.14). Here was the kairos surrounding that amphibole: in 480 B.C., Athens petitioned the Pythia for an oracle in the face of an impending Persian attack following the Battle of Thermopylae: What should it do? The Pythia's first response was horrible:

> Wretches, why do you linger here? Rather flee from your houses and city,
> Flee to the ends of the earth from the circle embattled of Athens!...
> Many a shrine of the gods will [the god of war] give to the flame for devouring;
> Sweating for fear they stand, and quaking for dread of the enemy,
> Running with gore are their roofs, foreseeing the stress of their sorrow;
> Therefore I bid you depart from the sanctuary.
> Have courage to lighten your evil. (Herodotus 7.140)

The Athenian delegation was terrified. It returned to the Pythia bearing olive branches and begged another prophecy. This time the Pythia said Athens would be saved by a "wooden wall." Back in Athens, the members of the Assembly scratched their heads over this second response: what did the god mean? Some argued that the thorny trees that once ringed the Acropolis were the "wooden wall" and that citizens should retreat there for their salvation. Themistocles countered this position by arguing that the "wooden wall" would be formed by the sides of Athens's ships, and that to survive, citizens had to leave the temples unguarded and engage the Persians in naval battle at Salamis. Eventually, Themistocles and his supporters won the debate, and the Persians were defeated (Herodotus 7.142–8.76).

Just as this amphibole required political dialogue to settle its meaning, so did the Pythia's riddles. Plutarch recorded perhaps the most famous of these in his *Moralia*.[56] The inhabitants of the island of Delos, Apollo's legendary birthplace, were suffering a plague, so they went to Delphi to ask the Pythia what they should do to stop it. The Pythia responded, "They should double the altar and then sacrifice on it." The task was not as simple as it seemed, for the altar was a cube, and so if they simply doubled each dimension, they would actually increase the volume of the altar eightfold, not twice. The Delians turned for help to Plato, who responded that there was no right answer, that the Delians had to work out the solution the best they could, for this was Apollo's design: "When the god gives out ambiguous oracles, he is promoting and organizing logical reasoning as indispensable for those who are to apprehend his meaning aright" (Plutarch, *Sign of Socrates* 386E–F).

These ambiguous Delphic pronouncements, when viewed through the lens of the Act ("What did the drama enact?"), highlight the classical antinomy between word (*logos*) and deed (*ergos*). As I mentioned earlier, *logos* indicated not just "words" in our present sense, but rather supported a whole constellation of meanings related to rationality and argument. However, the Greeks opposed *logos* to action just as Romans opposed the contemplative life to the active life. Socrates argued poignantly in his *Apology* for this distinction when he claimed the active life amounted to a death sentence for philosophers. He claimed that the story of his life justified that stance with "not mere words [*logous*], but what you honor more,—actions [*erga*]" (32a).

Delphi's amphiboles confounded the words/deeds antinomy. The mysterious references they contained could only be resolved through the political action of dialogue. We saw Herodotus acknowledge the dialogics of Delphi when he recorded how the Pythia criticized Croesus for not seeking clarification in the River Halys kairos (Herodotus 1.90). Such a clarifying dialogue did take place in the case of the "wooden wall" oracle; further, it continued in the Assembly between subpolities that supported different interpretations of the reference. And in fact, the dialogue itself never ceased; it simply morphed into other symbolic actions: the massing of ships, the planning of battle. But then the Athenian work of self-definition never ceased, either.

This special motivation that transcends the words/deeds antinomy, I call Prophecy, since it was the epitome of ancient prophetic ethos in Greek life. In fact, when Plutarch said Apollo gave an "ambiguous answer" in the Delian case, he was alluding to the pre-Socratic philosopher Heraclitus, who generations before had used Delphi to prove his point that everything was one because everything enfolded its opposite: "The Lord whose oracle is at Delphi neither speaks nor conceals but gives signs" (Heraclitus 18). If revelation and concealment are opposites, "giving a sign" entails both. This was the function of the Delian command—to prescribe a task whose execution entailed both words and deeds as the Delians debated with each other the correct geometry of doubling a cube.[57]

Scholars of prophetic practices across the ancient Mediterannean have recognized their dialogic nature.[58] Further, they have identified a telltale formula—the judgment formula—that frequently ritualized prophetic dialogue. Claus Westermann painstakingly reconstructed the pattern from Babylonian, Syrian, and Hebraic archives. It consists of an introduction in which the messenger states his right to speak, then a judgment, which can either be a restatement of covenant followed by a promise (isaiad) or, more commonly, an indictment of covenant transgression followed by a threat (jeremiad). Here is a brief illustration of the form from Jeremiah 14:14–15:

> Then the LORD said to me, "The prophets are prophesying falsehood in My name. I have neither sent them nor commanded them nor spoken to them; they are prophesying to you a false vision, divination, futility and the deception of their own minds.
>
> "Therefore thus says the LORD concerning the prophets who are prophesying in My name, although it was not I who sent them—yet they keep saying, 'There will be no sword or famine in this land'—by sword and famine those prophets shall meet their end!"

While this might seem at first blush to be a divine monologue (ventriloquized by the prophet), in fact, it frames a dialogue between Yahweh and the people, mediated by the prophet. During this dialogue, the symbolic actions each party takes condition the other parties' responses. The Old Testament is full of passages in which the people repent in response to a jeremiad. The prophet then petitions God for mercy, and God, in turn, rescinds the threatened punishment. And so the dialogue continues.

Rhetoricians recognize that such formulae—called "genres"—are triggered by recurrent kairoi; if a community or communities encounters the same class of crisis over and over again—such as natural disaster or foreign invasion—they predictably develop standard responses to it.[59] These genres constantly evolve as they interact with the constraints of each new kairos. But they have some persistent features can serve as entry points for scholars trying to recognize and understand recurrent kairoi. These genres are therefore important markers of the meme of prophetic ethos, along with the ancient Mediterranean judgment formula, amphiboles, and riddles. They include borrowed elements from wisdom literature, particularly aphorisms like Solomon's and parables like Jesus's.[60] What all these diverse genres have in common is that they are laconic.[61] They contain rhetorical holes—ambiguities, uncertainties, surrealities, and silences. These holes invite the audience to supply the missing meanings, the interpretations that tailor the prophecy to the present crisis. These invitations initiate the dialogue of Prophecy.

To sum up the special motivation of Prophecy at Delphi and elsewhere in the ancient Mediterranean: we have seen that it puts petitioners and prophets into dialogue; through this dialogue, the antinomy of words/deeds is transcended as the participants interpret, debate, and fill in the holes in prophetic genres.

Petitioners came to the Pythia with a certain idea of language—that it was an intellectual scalpel designed to carve up identity so that, laid out prone like the lobes of a sacrificial liver, it would serve monologue, tyranny, war, imperialism. What the Pythia gave Athens back in her prophecies was *logos* transformed—slippery, demanding, shot through with chaos. For her prophecies really said nothing at all until they came into the Assembly, where they goaded Athenians into dialogue with themselves and Others. The product of this dialogue was political certainty.

Conclusion: The Memetic Signature of Ancient Prophetic Ethos

In this chapter I used Burke's dramatic pentad to frame the following research question: How did prophecy intervene in the co-constitution of the best-known oracle and the best-known democracy in the classical world? After studying all the known forms of petitions to Delphi and checking results against other regional practices, I have identified the five key motivations of ancient prophetic ethos: Ascertainment, Authorization, Confirmation, Divination, and Prophecy.

To summarize the ethos constituted by the interaction of these five motivations: The ancient prophet helped a polity release itself from the horns of a dilemma by engaging it in a dialogue that reframed the dilemma in terms of covenant values; this provisional political certainty triggered political action that broadcast the polity's identity both back to itself and to Others.

Lest we think our need for such interventions fell away from us like the crumbling pillars of Apollo's temple, Delphic scholar Hugh Bowden reminds us how vulnerable our polities have remained to "acts of god":

> [E]pidemics were not understood until the nineteenth century; many climatic events are still beyond human understanding and control; 'chance' remains a recognised element in explaining the outcome of wars. A consequence of this is that when a community seeks a solution to such a problem from the gods it cannot test its correctness against human standards of "rationality." The answer has to be taken on trust.[62]

And whom do we trust to figure out epidemics and hurricane swarms? Scientists. The subtle substitution of scientific for prophetic ethos is the central curiosity that drives my project: I want to observe how the prophetic ethos was adopted and adapted by scientists as their polities increasingly tasked them with manufacturing certainty, first in the English Renaissance at the inception of modern scientific practices, then during the organization of science on an industrial scale, and finally at the present moment.

In each of these kairoi to come, I will use the five motivations of prophetic ethos as sightlines to observe both how it persisted and how it mutated in its crossings with scientific ethos—and how the hybrid prophetic-scientific ethos might explain the ethical problems our science advisers face today. Like the Delphic oracle, my topical genealogy will not yield absolute knowledge. After we are done, the bully pulpit of scientific-prophetic ethos will not be fully determined. But I believe it will be illuminated.

3 THE NATURAL MAGICIAN AND THE PROPHET

FRANCIS BACON'S ETHICAL ALCHEMY

Picture a windy day in early seventeenth-century London with clouds and sun streaming together over the smoky rooftops. The burnt tower of Old St. Paul's still soars over this city of some 200,000 souls despite the loss of its prodigious steeple, which rained molten lead into the cathedral when lightning struck it in 1561 (an omen that Protestants and Catholics alike interpreted as God's judgment on their denominational rivals). Something or someone now moves at the top of the ruined tower. If we could sweep closer like a bird, we would see a man teetering on boards balanced across the yawning opening. In his hands are a magnet and a needle. He is checking to see if the magnet's force is diminished here, high above the ground believed to generate its power. It isn't. The man hurriedly tucks his experiment into a pocket and scrambles down from the tower.

We are not certain who this man was, but his experiment appears in Francis Bacon's *Inquiry Respecting the Magnet*, designed for inclusion in *The Great Instauration* (1620), Bacon's *magnum opus* on natural philosophy. Given Bacon's health around this time, it is likely that the man trembling on the tower scaffolding was not Bacon but his amanuensis, William Rawley.[1] But it doesn't matter; it could have been any one of the growing community of London experimentalists. Forty years later, Robert Hooke would be up there stringing pendulums for Robert Boyle; this was one of the last St. Paul's experiments before the Great Fire of London in 1666 finished it (and with it, luckily, the plague of 1665) and the rubble was cleared to make way for Christopher Wren's neoclassical design.[2]

The Old St. Paul's tower experiments dramatize the story I tell in this chapter. It is a story about how seventeenth-century experimental philosophers looking for a stable ethos in early modern society first adopted and adapted the political platform offered by prophetic ethos. And Bacon is the perfect figure to begin the story with. More specifically, I examine his grand ethical alchemy, the elegant rhetorical experiment in which he alloyed the ethē of the magician and the prophet to yield an argument that good Christian gentlemen owed it to their country to devote themselves as much to the book of God's Works as to his Word. This alchemy is our business in this chapter; the next chapter explains how the Royal Society built on the foundation of Bacon's prophetic legacy to establish experimental philosophy as the new civic oracle of England—and the forerunner of our "science."

Francis Bacon

As part of a collection of elegies written on the occasion of Bacon's death in 1626, the poet "R.P." exhorted his spirit, "Burst be thy prison foundations, seek thou the royal Jacobus / Showing that loyalty follows, even the true soul's hereafter. / Then, from Law's tripod announcing, shalt thou thine oracles utter / To the disciples of Themis."[3]

This fragment encapsulates so much of Bacon's tumultuous career. Born to wealth and privilege—the son of Elizabeth I's Lord Keeper, Nicholas Bacon—Francis had good reason to expect a course like his father's. But perhaps his birth in the year that lightning struck St. Paul's was an omen: a series of major reversals of fortune, missteps at court, and grave financial miscalculations put Bacon's career—and sometimes his neck—at risk throughout his life. He entered the bar at Gray's Inn and scrambled to meet his debts by piecing together various legal and advisory assignments, grumbling all the while that what he really wanted was the contemplative life of the natural philosopher.[4] He tirelessly pressed every advantage he had at court; he was a member of Elizabeth I's privy council for a while, then James I's attorney general. When he finally clawed his way up to Lord Chancellor—the highest civilian position in the realm—and James created him Lord Verulam and Viscount St. Albans shortly after, Bacon clearly thought he had arrived. But he did not get to enjoy the view from his plinth for very long before charges of bribe taking brought him low. Although he spent only two days in the Tower, he exited it a ruined man—ruined, but still faithful to the king who had most likely betrayed him.[5] He spent most of the rest of his life out of the public eye and died deeply in debt. And yet, in spite of all his setbacks and laments, Bacon left the seventeenth century what amounted to one of the most significant English contributions to natural philosophy, including the famous *Advancement of Learning* (1605), the *New Organon* (1620), the *Sylva Sylvarum* ("Collection of Collections" [1627]), and the popular fable the *New Atlantis* (1627).

While Bacon's tragic biography is intriguing enough on its own to fill this book, our focus, trained as it is on all things prophetic, returns to the funereal ode—to two details of it in particular: First, the poet compares Bacon explicitly to the Delphic oracle; but second and more importantly, the Pythia's tripod has become "Law's tripod." In this poet's vision, the English state is the god at Delphi, and Bacon is her prophet. Lest we think these oracular metaphors were one poet's passing fancy, consider Ben Jonson's contribution to Bacon's 60th birthday festivities: "Hail, happy genius of this ancient pile! / How comes it all things so about thee smile? / The fire, the wine, the men! and in the midst / Thou stand'st as if some mystery thou didst!"[6] And we hardly have to struggle to find the prophetic metaphors in this paean: "Bacon, like Moses, led us forth at last, / The barren wilderness he past, / Did on the very border stand / Of the blest promis'd Land, And from the Mountains Top of his Exalted Wit, / saw it himself, and shew'd us it."[7]

This last was Abraham Cowley's prefatory ode to the *History* of the fledgling Royal Society, published 41 years after Bacon's death. It is this constitution of Bacon as the prophet of the new experimental philosophy, and of the state, that I examine in this chapter.

It might seem that pulling this genealogical thread would mean demonstrating, first, that Bacon made a hybrid of experimental-philosophic and prophetic ethē and, second, that the founders of the Royal Society received this hybrid ethos as their model and Bacon as their prophet. But our dramatistic focus reminds us that a complicated rhetorical function like ethos can never be just one man's invention. Accordingly, we have to look at the wider kairos surrounding Bacon's philosophical arguments to appreciate the constitution of experimental-prophetic ethos.

In following this line of inquiry, I am agreeing in outline with Thomas Lessl's argument that Bacon created an ethos for experimental philosophy in the seventeenth century by cobbling together existing ethē.[8] Experimental philosophy did not have a place in English society: no one recognized it as playing a crucial role, and thus it was assigned neither respect nor public resources. Bacon and the other experimental philosophers of his day—such as William Harvey, William Gilbert, and Hugh Plat[9]—could not despite their considerable material resources conjure such an ethical space out of thin air. While money and power certainly help support an ethos, it is fundamentally a rhetorical role and thus has to be argued into existence starting from recognized and accepted models. Lessl has argued that the primary model was the Protestant reformer. I would add to the recipe the powerful but unstable ethos of the natural magician. In the remainder of this chapter I detail the ethical alchemy by which Bacon took these rhetorical resources and transmuted them in his writings into a new experimental-prophetic ethos.

Studying Bacon

In considering Bacon's contributions to the development of scientific ethos, I am pulling a thread in what is indeed a well-worn tapestry. Bacon studies have had a series of epochs: in the nineteenth century, he was the celebrated author of modern science; to early twentieth-century positivists, he was a bungler who managed, if he managed anything at all, to set the development of proper science back a hundred years; late twentieth-century historians have resuscitated him primarily as a channel into some of the movements he participated in, such as the neo-Platonic philosophy of Forms, the plain speech movement, and utopianism.[10]

Let me clarify where I stand against this backdrop. I do not think Bacon single-handedly invented anything, not even the ethos of the science adviser. Nor do I argue that history selected him by virtue of his considerable eloquence and social power to set English society on the path toward a new civic epistemology.[11] But I do agree with recent scholars that Bacon serves as an excellent entry point into arguments that developed during his lifetime about the future of the English state and how to prosper it. Bacon was by all contemporary accounts extraordinarily sensitive to these arguments, and he did manage to consolidate them into a coherent vision that was taken up and elaborated during a sympathetic kairos a half century later. In this way, like the Delphic oracle, Bacon was a prophet.

This, he would have told you himself. In both his legal work and his philosophical writings, Bacon frequently referred to himself in prophetic terms, as a "priest" of nature who interpreted its "oracles," read its "omens," performed natural "auguries," and provided an "apocalypse, or true vision" of its order.[12] And indeed, although Bacon outlined a new experimental method in *New Organon*, he performed very few experiments himself. Instead, he dedicated himself to calling his peers to reevaluate and reprioritize natural philosophy with respect to their sworn duties to God and Crown. In the process of scrutinizing this argument and its legacy, I hope to shed new light on a few issues that still concern Bacon scholars, among them: the rhetorical resources Bacon drew from in helping to constitute the experimental ethos, the origins of his inductive method, and the apparent contradictions in his recommendations for the style in which natural philosophy should be written.

Bacon's Projects

Bacon's interwoven projects were the reformation of natural philosophy toward "operation," productive knowledge that would benefit the common good, and the creation of a powerful and stable ethos in English society for the man who pursued this philosophic work (namely, himself). On the first point Bacon wrote obsessively during the 1590s and early 1600s. The most polished formulation of the argument appeared in *Advancement of Learning* (1605), the upshot of which was that all previous approaches to natural philosophy had made various grave errors that prevented the achievement of its proper goal—an operative knowledge of Nature by which philosophers could repair the damage done to Nature in the Fall from Eden. Such reparations would usher humankind into a New Jerusalem on earth. These glorious ends were achievable only through systematic experiments aimed at inducing the Forms that underlay and shaped all objects in Nature. Once discovered, these Forms could be manipulated by philosophers to repair or create objects.

But the success of Bacon's program depended crucially on the gentlemen who had the leisure, education, and resources to pursue it; more to the point, it depended on their monarch. Bacon aimed at no less than the instauration, or rebirth, of English society as a whole. A project of this magnitude had to have the blessing (not to mention the funding) of the Crown. Bacon started working on Elizabeth I in the 1590s. Their relationship was so mercurial that it is uncertain how far he was able to press his case, but a script he wrote for the annual Gray's Inn revels reveals the powerful role in England's future he imagined for the experimental philosopher. The revels assembled the students of Gray's Inn as a microcosm of English society, including the court and a mock "Prince of Purpoole." The Queen herself was the Inn's patron and often sent representatives to the revels' Grand Nights. Bacon was clearly counting on their attendance on January 3, 1594/95, because into the mouth of the second of the prince's six "councilors," Bacon put an exhortation to the mock prince to become a natural philosopher. The councilor admonished "that you bend the excellency of your spirits to the searching-out, inventing, and discovering of all whatsoever is hid and secret in the world; that your Excellency be not as a lamp that shineth to others and yet seeth not itself, but as the Eye of the World that both carrieth and useth light." He next reminded the prince of the ancient sympathy among power, philosophy, and magic:

> Amongst the Persians, the kings were attended on by the Magi. The Gymnosophists had all the government under the princes of Asia;

> and generally those kingdoms were accounted most happy, that had rulers most addicted to philosophy. The Ptolemies in Egypt may be for instance; and Salomon was a man so seen in the universality of nature that he wrote an herbal of all that was green upon the earth. (8.334)

The councilor finished by begging the prince to outfit himself with a library, an experimental garden, a curiosity cabinet (where Renaissance aristocrats kept exotic specimens), and a laboratory.

Whatever version of this plan might have been carried back to Elizabeth, she evinced no interest in furthering Bacon's designs. Bacon tried harder with James I, dedicating several of his books to the king and making sure he was presented with beautifully bound copies. In the dedicatory essay to *The Great Instauration* (1620), Bacon presented himself as an experimental philosopher:

> And thus I conceive that I perform the office of a true priest of the sense from which all knowledge in nature must be sought unless men mean to go mad, and a not unskilful interpreter of its oracles, and that while others only profess to uphold and cultivate the sense I do so in fact. Such then are the provisions I make for finding the genuine light of nature and kindling and bringing it to bear. (4.26)

The prophetic ethos Bacon claimed here is striking; so is his promise to bring his arcane knowledge "to bear" on the prosperity of England. It was a tantalizing prospect for the monarch, who apparently read—or tried to read—*The Great Instauration*; he quipped to one of his courtiers that the methodological section, the *New Organon*, was "like the peace of God, that passeth all understanding."[13] In any event, Bacon never received serious assistance in his projects from James.

These efforts might have failed, but only part of the work of establishing the new experimental ethos was political and financial: a great deal of it was rhetorical. And as we glimpsed in the praise heaped on Bacon at his death and afterward, the hybrid ethos he constructed in his writings—part prophet, part court magician—was eagerly taken up by future generations of experimenters. I will detail how Bacon constructed this hybrid ethos in his own writings, but first we need to understand the ethical material he had to work with, particularly the ethos of the natural magician, a role unfamiliar to us but one that served as the touchstone of Bacon's ethical alchemy.

The Operative Ethos of Natural Magic

Although there is no doubt that Bacon drew heavily on prophetic ethos in creating the "experimental philosopher," this role had to produce wonders and technologies that could be "brought to bear" in remedying England's problems. Such miraculous works were not to be found among the prophets in Bacon's day; they were found in the spectacular, powerful, and unstable figure of the natural magician. Natural magic, which today we know primarily through its central discipline, alchemy, produced—more often by accident than design—magnets, explosions, predictions of eclipses, and the distillation of powerful acids. These are all works that today we would assign to science or technology.[14] The writings of natural magicians, or Hermeticists, were intensely popular in Bacon's day. In addition to the predictions of Nostradamus, these included the prophecy and wisdom literature of Pico della Mirandola, Marsilio Ficino, Cornelius Agrippa, Giordano Bruno, Giambattista della Porta, Gerolamo Cardano, and Paracelsus.[15]

Ficino traced the roots of Hermeticism back to a legendary predecessor of Plato named Hermes Trismegistus, "thrice great," referring to his hybrid role as king, philosopher, and priest.[16] Trismegistus had purportedly maintained that humans had fallen from a demi-godlike state but could be restored to it if God's knowledge of Forms were reassembled. These Forms were a great deal like Plato's ideal Forms, but Ficino's twist on them made them operative: they were God's recipes. When Christians were reunited with God through Christ's sacrifice, they regained partial knowledge of these Forms. But this knowledge could be amplified through devout study, prayer, and experimentation until it reached a level where it could be used to actually replicate and repair matter. In fact, Ficino and others argued, it was humankind's god-given duty as the "interpreters of nature" to comb the Book of Nature for fragments of the Forms and reconstitute them.[17] In short, Christ had provided humankind with the instruction manual for Eden, and now it was the job of inquisitive Christian gentlemen to perform the assembly. This was the basic restorative logic that underlay all the popular theories of natural magic in England.

Natural magicians enjoyed extraordinary political power and success. In addition to the popularity of their writings, some, such as Bruno and John Dee, became court magicians. But the occult nature of their teachings made their positions extraordinarily unstable. Bruno was burned at the stake in 1600 for, among other things, declaring himself a prophet in his treatise *Ash Wednesday Supper*.[18] Cornelius Agrippa was never imprisoned but recanted his belief in magic under significant pressure several years before he died in 1535.

In England, Dee epitomized the unstable nexus among magical, religious, and scientific rhetoric during the sixteenth century. A gifted mathematician, astronomist, and Hermeticist, Dee swore that he had been chosen to magically reconcile the schismed church and usher in a new Eden with Elizabeth I as its holy monarch.[19] These flattering claims won him a position as Elizabeth's chief astrologist. But then Dee embarked on an obsessive campaign to communicate with and control angelic powers via a language of geometric symbols. Elizabeth protected him from serious censure for these occult activities during her life, but James I disciplined Dee with the 1604 Act Against Conjuration. When Dee died in 1608, his unparalleled library of natural philosophy and magic had been stripped to remainders to meet his debts.

Ryan Stark has argued that in proposing his philosophic reforms, Bacon utterly rejected the natural magic tradition that had motivated Dee's rise and fall.[20] But to make this categorical statement, Stark has had to discount a substantial thread in Bacon scholarship and in many of Bacon's own writings. Although it is undeniable that the occult orientation and disordered methods of natural magicians came in for sharp criticism in the *Advancement*, Bacon's admiration for the more systematic and productive parts of the tradition is clear in that document and well attested elsewhere.[21] Inducing and manipulating the Forms underlying nature is the very signature of Bacon's proposed method in *The New Organon*, as we will see shortly. Scholars have identified the source of his theory of Forms as Ficino; further, Bacon's conception of how Forms operated in nature comes from various Renaissance alchemists.[22] In addition, his only substantial contribution toward the comprehensive natural history he had proposed, the *Sylva Sylvarum*, was substantially cribbed from Cardano and della Porta.[23]

Although the operative ethos of the natural magician was pivotal for Bacon's proposed reforms, he did need to purify it of its occult strains; otherwise, it would not endure as a stable ethical stance for Reformation experimenters. The most natural candidate for a purifying spirit was the cultic Protestant prophetic ethos. This ethos, based on classical prophetic ethos with some important mutations, was socially current during Bacon's day to the degree that speakers could prophesy in Parliament.[24] More importantly, millennialist Protestants had embraced aspects of natural magic during Edward's and Elizabeth's reigns (and would again during the Interregnum) as furthering their mission to reform English society into a New Jerusalem. Thus, the Protestant prophetic ethos was an organic purifying and stabilizing agent for Bacon's experimental ethos.

Prophecy in Late Tudor and Elizabethan England

The king who threw Bacon into the Tower of London also sponsored the most famous English translation of the Bible. But it had been available in English long before James's translation; and in Latin, even longer. From the stories of Elijah, Jeremiah, Moses, and Paul, Bacon's contemporaries learned to recognize the markers of classical prophetic ethos. In addition, they and even their fathers would have read Latin accounts of the Delphic oracle in school, as well as Socrates's reflections on *mania/manteia* and the mythologies surrounding the prophetic legacies of Prometheus and Hermes.[25] These understandings of ancient biblical and pagan prophecy would have been filtered, however, through an important native prophetic movement.

The political upheaval following Henry VIII's Reformation of the Church in England spawned a millennialist Protestant rhetoric that reinterpreted the book of Revelation as describing current times: the expulsion of Catholicism from England, like the defeat of the anti-Christ in the Bible, was interpreted as the harbinger of the Millennium, Christ's thousand-year reign on earth that would establish the New Jerusalem and repair the damages wrought by the Fall and by the wicked pagan and Catholic regimes that had until recently oppressed Creation.[26]

In the absence of the mediating authority of Catholic clergy, it fell to congregations to perform this millennialist hermeneutics. Ad hoc groups of Protestants, men and women alike, started gathering to interpret prophetic passages from the books of Daniel, Ezekiel, and Revelation.[27] But they had to perform this ecumenical divination somewhat privately, as it opened them to criticisms of "enthusiasm," a pejorative referring to the heretical and anarchical privileging of individual revelation above the received authority of the Anglican Church. To reassure these bishops, key Protestant prophets such as Robert Crowley and John Bale argued that their gatherings served Edward VI, Henry's nine-year-old heir, whom they cast as the prince of the New Jerusalem.[28] By tying prophetic authority to royal authority in this way, Crowley and Bale restored to the prophetic ethos the prestige it lost in the disruption of the Catholic hierarchy.[29]

Patriotism was thus a confirming sign of the new hybrid classical-Protestant ethos in Bacon's day. What were the other signs? We can look for answers in William Perkins's influential contemporary treatise *The Art of Prophecying*. The method directed the preacher, "1. To read the Text distinctly out of the Canonicall Scriptures. 2. To give the sense and understanding of it being read, by the Scripture it selfe. 3. To collect a few and profitable points of doctrine out of the naturall sense. 4. To applie (if he have the gift) the doctrines rightly collected to the life and manners of men, in a simple and plaine speech."[30]

What did a prophetic sermon preached by this method sound like? Thomas Lessl has given us an example from Bacon himself. It appears in the section of the *Advancement* in which Bacon assembles his key argument that it is the Christian duty of English gentlemen to pursue experimental philosophy. Bacon begins by quoting Ecclesiastes 3:11: "God hath made all things beautiful, or decent, in the true return of their seasons: Also he hath placed the world in man's heart, yet cannot man find out the work which God worketh from the beginning to the end."[31] Bacon then interprets this passage according to Perkins's method. First, he "gives the sense" of the verse and related ones from the biblical wisdom literature. Then, he "collects" from these texts two main points: (a) creation is essentially good, since God made it and God is good, and (b) it is not only good for philosophers to try to induce the laws by which God made nature, it is their Christian duty. Finally, Bacon "applies" these principles by arguing that Christian gentlemen can help to correct the defects of the Fall—disease, poverty, early mortality, and so on—if they reform philosophy as he has suggested. However, they must be careful to avoid the "three vanities" of learning: personal pride, covetousness, and Satanic pride.[32] The best safeguards against these sins are (a) to subject individual reason to the "oracle of God's works," namely, observations of nature,[33] and (b) to make the goal of philosophy Christian charity, "the good of men and mankind."[34]

So, as Lessl points out, Bacon positions himself as a reforming prophet by divining political applications from sacred texts. Lessl does not mention this, but while Bacon was drafting the *Advancement*, he also wrote an advisory document for James I in which he commended the corporate discussion and testing of prophecies by church elders as safeguards on the practice (10.119). If we take this advice together with Bacon's sermon, Perkins's method, and what we know about millennialist prophecy in Bacon's day, we can descry the outline of the Protestant prophetic ethos. In addition to performing *loyalty to the Crown*, the new prophetic ethos was *systematic and corporate*: it proceeded by a prescribed method, and truth emerged only from the group's validation of any proffered interpretation. This political epistemology formalized the dialogic tendency in the classical meme.

In addition to these markers of patriotism and democracy, Protestant prophecy *aimed at spiritual and political restoration*.[35] We see here a strong echo of the classical judgment formula, with an emphasis on promises rather than threats.

This, then, was the dominant prophetic ethos in Bacon's London. The connections between the teachings of the Hermeticists and the Protestant prophetic ethos should be coming clear at this point—particularly the shared emphasis on divination from sacred texts (the "two books" of the Word and the World) as well as on millennialist restoration. The two ethē were in many ways natural political allies. Natural magicians could provide control over nature, and millennialist Protestants considered domination over nature a hallmark of the New Jerusalem.[36] Bacon exploited this natural sympathy to alloy a new ethos—the experimental philosopher.

Francis Bacon's Ethical Alchemy

This was the ethical alchemy by which Bacon created the new ethos of the experimental philosopher: first, he took the traditional natural philosopher's ethos and smelted out all the Scholastic wrangling, heretical sorcery, and experiments "too silly to be random."[37] What he had left after this initial purification were the productive methods of the best-respected natural magicians. To this ethos Bacon added the "spirit" of the Protestant prophet to digest it of its occult tendencies and to fix its polarity firmly pro bono publico, toward the good of the English state.[38]

Bacon did not perform this alchemy all at once or in one place: the transmutation took 20 years and is scattered throughout his writings on experimental philosophy. For coherence's sake, I recollect it here under the topoi I established at the end of chapter 2: Ascertainment (the purpose of consulting prophets); Authorization (the process of establishing an oracle); Confirmation (recognizing signs of the prophet's privileged access to knowledge); Divination (the method for reading/writing that knowledge); and Prophecy (the dialogue between prophet and polity).

Ascertainment: Knowing by Doing

Bacon opened *The New Organon,* his manifesto for his new experimental method by saying, "Now my method though hard to practise is easy to explain and it is this: I propose to establish progressive stages of certainty" (4.40). By "certainty" Bacon meant something specific: first, the experimental philosopher would systematically divine the "alphabet" of Forms by which God had originally written the Book of Nature;[39] then, the philosopher would test his knowledge by spelling out his own new works of creation.[40]

This fruitful knowledge was called "maker's knowledge" in the natural magic tradition, and its purpose was reacquiring humanity's lost dominion over creation, which would in turn restore the world to prelapsarian perfection.[41] In Bacon's Gray's Inn speeches and in *The New Atlantis,* he argued that the experimental philosopher was the one to accomplish this restoration for England. Bacon argued in a similar key in *The New Organon*, "For man by the Fall fell at the same time from his state of innocence and from his dominion over creation. Both of these losses, however, can even in this life be in some part repaired; the former by religion and faith, the latter by arts and sciences" (4.248). The segregation of religious and philosophical spheres expressed in this passage was pivotal to Bacon's plan for authorizing experimental philosophy as a civic oracle.

Authorization: Salomon's House

Bacon's most complete model for how he envisioned the ethos of the experimental philosopher operating with respect to the English state appears in the *New Atlantis*, his wildly popular fable about an ideal scientific society.[42] His narrator and companions are English merchants cast up like the prophet Jonah on the shores of the fictional island nation of Bensalem.[43] They are nursed back to health by the Bensalemite doctors, and it is not long before the voyagers learn the source of their hosts' quasi-miraculous skills: Salomon's House, or the College of the Six Days Work.

Many of the key features of Salomon's House parallel the features of ancient civic oracles. First, it has been established for the purpose of "finding out of the true nature of all things" and the "causes, and secret motions of things."[44] Second, it has been established outside the polity, such that a visit from a Father of Salomon's House is a rare and auspicious occasion in the city.[45] One of these Fathers proves to be the most powerful person the narrator meets in Bensalem, yet this elder comports himself with humility toward the narrator and his companions. The Fathers conduct their work in secrecy, only sharing with the Bensalemites what they feel to be beneficial for the common good.[46] In fact, they employ a great number of "hermits" who communicate observations from remote towers and underground caves.[47] Finally, the chief political responsibility of Salomon's House is to "declare natural divinations of diseases, plagues, swarms of hurtful creatures, scarcity, tempests, earthquakes, great inundations, comets, temperature of the year, and divers other things; and [to] give counsel thereupon what the people shall do for the prevention and remedy of them."[48]

We would recognize much of the work of Salomon's House today as science, but it is also clearly and strongly prophetic. The classical markers are obvious: asceticism, humility, the leveraging of privileged access to knowledge to engage the polity in dialogue about its covenant values, its independence from politics as usual. Another classical aspect carried forward here that might not be immediately apparent is the way Salomon's House projects voices from the margins of Bensalem into its center. It is undeniable that the Fathers themselves are privileged in their dress and lifestyle—just as Bacon and his fellow gentlemen were in English society. It is important to remember that though the middle class was growing in Bacon's day, and social mobility and literacy were increasing, the men who had any say in how England was governed amounted to 5 percent of the population, at the outside.[49] These were Bacon's recruits. And yet the concern the Fathers of Salomon's House exhibit for the poor and for the universal health and happiness of Bensalem dramatizes Bacon's argument in the *Advancement* that the proper aim of experimental philosophy is restorative charity. In this way Salomon's House carries on the classical prophetic tradition of speaking suffering to privilege.[50]

But the prophecy Salomon's House engages in has strongly Protestant hallmarks as well: for instance, natural truths are divined systematically and corporately, and the Fathers are preoccupied with applications to agriculture, industry, and defense.[51] The *New Atlantis* amplifies the millennialist prophetic strain in Bacon's philosophy to a near-fever pitch.[52]

But most importantly, by establishing Salomon's House as separate from the Church in Bensalem and endowing it with its own prophetic figures (starting with its figurehead, the ancient Hebrew prophet-king Solomon), Bacon dramatizes the pivotal argument in his ethical alchemy: natural philosophy needed its own prophets.

As Lessl and numerous others have already argued, Bacon made natural philosophy "safe" for Christian gentlemen to pursue by disengaging it from the divination of God's will. It would reveal nothing about God's mind or sovereign will—this field was left to theology.[53] This rhetorical decoupling of theology and natural philosophy—technically quite provisional in Bacon's writings[54]—has been identified as crucial to the shift from medieval to modern thought.[55] Bacon was hardly the first to lay out these parallel paths to divine and natural truth, as we witnessed in Ficino's "two books" arguments above.[56] But Bacon was the first, with the *New Atlantis,* to flesh out the ethical analogy between the clergy serving as God's prophets and the experimental philosophers serving as Nature's prophets.[57] Bacon's arguments, and his own confirming performances as a "priest of Nature," help revise our understanding of the Renaissance "two books" discourse as ethically driven—aimed at crafting a stable ethos for the men who wanted to pursue experimental philosophy independently from church and court.

Confirmation: Digging the Clay

In his writings on experimental philosophy, Bacon provided a glimpse of what a "priest of Nature" should look like. Here is the invocation that opens a precursor essay to the *Advancement*, titled *The Masculine Birth of Time*:

> God the Word, God the Spirit, we pour out our humble and burning prayers, that mindful of the miseries of the human race and this our mortal pilgrimage in which we wear out evil days and few, they would send down upon us new streams from the fountains of their mercy for the relief of our distress; and this too we would ask, that our human interests may not stand in the way of the divine, nor from the unlocking of the paths of sense and the enkindling of a greater light in nature than may any unbelief or darkness arise in our minds to shut out the knowledge of the divine mysteries; but rather that the intellect made clean and pure from all vain fancies, and subjecting itself in voluntary submission to the divine oracles, may render to faith the things that belong to faith.[58]

Here we see echoes of the prophetic judgment formula (in the indictment of "vain fancies" and the promise of "relief of our distress") as well as asceticism, humility, and compulsion. Further, Bacon presented himself as engaged in righteous battle against the false prophets leading English minds astray. Accordingly, he went on in the *Birth*, as he would in the *Advancement*, to denounce previous natural philosophers. Against Hippocrates he railed,

> [H]e snatches up a few idols—not the monstrous idols of the great speculative thinkers, but a slim and elegant variety which haunts the surface of science. These he swallows, and swollen with this diet, half scientist and half sophist, protecting himself according to the fashion of his age by an oracular brevity, after long delay he brings out a few maxims, which Galen and Paracelsus take for oracles and quarrel with one another for the honour of interpreting. But in truth the oracle is dumb.[59]

Bacon pleaded with his readers as his prophetic "sons" to carry the flame of the true method he alone advocated:

> [T]he present is like a seer with two faces, one looking towards the future, the other towards the past. Accordingly I have decided to

> prepare for your instruction tables of both ages, containing not only the past course and progress of science but also anticipations of things to come. The nature of these tables you could not conjecture before you see them. A genuine anticipation of them is beyond your scope, nor would you even be aware of the lack of it unless it were put into your hands. It is a compliment reserved to some of the choicer spirits among you whom I hope to win thereby.[60]

The "tables" are probably the taxonomy of knowledge Bacon published in the *Advancement*. Bacon lamented the exhausting work that his program required. His secretary, William Rawley, reported in the introduction to the *Sylva Sylvarum*

> [t]hat [Bacon] (who thinketh hee deserveth to be an Architect in this building,) should be forced to be a Work-man and a Labourer; And to digge the Clay and burne the Brick; And more than that, (according to the hard Condition of the *Israelites* at the latter end) to gather the Strawe and Stubble, over all the Fields, to burn the Bricks withall. For he knoweth, that except hee doe it, nothing will be done: Men are so sett to despise the Meanes of their owne good.[61]

In all these performances Bacon imitates the ethical models of the prophets and saints of the Bible—but with an important twist. He is not suffering for the Gospel but for experimental and operative knowledge. The extraordinary labors he complains about to Rawley are imposed by the rigors of the new method he has designed.

Divination: The Inductive Method

As I mentioned briefly above, Bacon admired Protestant prophecy for its systematic, fundamental, corporate, and progressive divinatory method. This method shared its hermeneutic focus and millennialist motivation with natural magic. But its emphasis on group validation could counteract the unfortunate tendency of natural magicians to closet themselves in their laboratories, melt things willy-nilly, and then conclude that angels were speaking to them in squares and triangles. In his methodological manifesto, the *New Organon*, Bacon applied the Protestant divinatory method to the rational achievement of maker's knowledge.[62] Previous historians and philosophers of science have overlooked this particular influence on the development of Bacon's experimental method, preferring to locate its origins in Stoic philosophy, Ramist dialectic, Hermeticism, or the methods of contemporary experimenters.[63] As I argued in chapter 1, genealogies are tangled affairs, and so in arguing that Bacon derived his method by analogy from the Protestant divinatory method, I am not seeking to cut off these previous accounts. Rather, I am following a thread that I believe tells a convincing story about not only the method's structure but also its political function—establishing an ethos that shadowed the established and respected ethos of the Protestant prophet.

In the *New Organon,* Bacon laid out his new method in a series of Delphic-sounding aphorisms, such as the famous third: "Human knowledge and human power meet in one; for where the cause is not known the effect cannot be produced. Nature to be commanded must be obeyed; and that which in contemplation is as the cause is in operation as the rule" (4.47). In this section of the *New Organon* we can see the fundamentalism of Bacon's method—its obsessive devotion to the Book of Nature—as well as its insistence on systematic induction, and its operative end.

Bacon followed up this prophetic series with an equally famous catalog of four false idols that had to be cast from the temple of the human intellect before his new method could properly take hold: the Idols of the Tribe (perception biases), the Cave (personal bias), the Marketplace (limitations of language, particularly metaphor), and the Theater (preconceptions based on previous philosophies, particularly Aristotle's) (4.53–54). Once again, Bacon recommended corporate validation, hermeneutic fidelity to the Book of Nature, and operation as correctives that would banish these idols.

What exactly was the right philosophical method, then? Bacon insisted that mere natural observation was not enough; the philosopher had to "vex" nature into giving up her secrets.[64] Bacon had opened this theme previously in the *Wisdom of the Ancients.* In his gloss on "Proteus, or Prophecy," Bacon reminded readers that this prophet of Neptune would only reveal what he knew about past, present and future if his petitioner were able to bind him and then endure all his horrible transformations. "The sense of this fable relates, it would seem, to the secrets of nature and the conditions of matter," Bacon argued (6.117). If the natural philosopher "vexed" nature via experiments, matter would reveal its true composition. "And whereas it is added in the fable that Proteus was a prophet and knew the three times; this agrees well with the nature of matter: for if a man knew the conditions, affections, and processes of matter, he would certainly comprehend the sum and general issue (for I do not say that his knowledge would extend to the parts and singularities) of all things past, present, and to come" (6.118). In short, by applying the correct divinatory methods, natural philosophers could become natural prophets.

These were the correct methods: one group of workers was to vex a natural phenomenon or "nature" in different experiments aimed at isolating the operation of the Form (formal cause) of the nature; the results of these experiments were recorded. Then, a different group of assistants would classify the results using a system related to the ancient inductive system of *genus differentia* (noting familial similarities and differences).[65] By examining this tabulation, the natural philosopher could winnow out any conditions unnecessary to produce the nature, at which point "there will remain at the bottom, all light opinions vanishing into smoke, a Form affirmative, solid and true and well defined" (4.146). The Forms thus discovered could then be recombined to reproduce fire or any other nature that could effect "an improvement in man's estate and an enlargement of his power over nature" (4.247). This method mirrored Protestant divination of principles for moral application—but with the Book of Nature substituting the Bible as the community's textual touchstone.

Bacon prescribed rhetorical methods for the sharing of the knowledge discovered by this divinatory method: chiefly, plain style; fable or "enigmatic" method; and aphorism. Previous scholars have considered these recommendations to be contradictory, but when we put them in the context of prophetic ethos, they emerge as a coherent rhetorical strategy.

Prophecy: Imaginative and Plain Style

Bacon's famous advice about "plain style" in reporting natural knowledge has been hotly debated: the contentions are that first, his "plain" style was not all that plain by modern standards; second, he apparently contradicted his "plain style" advice by concurrently recommending two "imaginative" methods—fable and aphorism—for the reporting of natural knowledge.

Bacon traced his imaginative methods back to ancient wisdom literature, particularly to Solomon: this genealogy reminds us that fable and aphorism were traditional genres of classical prophecy.[66] Bacon thought these genres ideal for explicating natural philosophy because by hinting rather than demonstrating, they extended an invitation to debate and further investigation. They also performed humility via their refusal to directly state truths.[67]

Bacon employed both genres in his own writings on natural philosophy. The *New Organon* and the *Sylva Sylvarum* contain extended passages of aphorisms. As for fables, we have the famous *New Atlantis*. Bacon's popular *Wisdom of the Ancients*, in which Proteus appeared, constituted an extended attempt to generate aphorisms for life from ancient fables. In the introduction to the latter work, Bacon elaborated on the seemingly contrary use of parables (a term he used interchangeably with "fables"), "to disguise and veil the meaning, and . . . to clear and throw light upon it" (6.79). He wished to concentrate on the second use:

> I mean the employment of parables as a method of teaching, whereby inventions that are new and abstruse and remote from vulgar opinions may find an easier passage to the understanding . . . For as hieroglyphics came before letters, so parables came before arguments. And even now if any one wish to let new light on any subject into men's minds, and that without offence or harshness, he must still go the same way and call in the aid of similitudes. (6.80)

This passage does appear to contradict his earlier recommendations to use fables to hide difficult or dangerous natural knowledge from the uninitiated. We should note, however, that Bacon is applying the fable to two different kinds of knowledge in the two instances. In the *Advancement* he treats the communication of natural knowledge to fellow workers; by contrast, in *Wisdom* he induces metaphilosophical aphorisms suggesting how science should be prosecuted and evaluated in Jacobian England. If Bacon wanted the secrets of nature to circulate only among the initiated, he nonetheless needed the rest of the polity to value natural philosophy highly enough to support it politically and financially. Everyone needed to stay in dialogue to ensure the success of Bacon's scientific program, and he found the "enigmatic methods" an effective rhetorical strategy to keep the dialogue going.[68]

So much for the imaginative methods: it is Bacon's recommendations for plain style that have riled up historians of rhetoric, who have reached assorted contradictory conclusions.[69] The brief passage that has generated most of the debate is found in the *Preparative Towards a Natural and Experimental History*, published with the *New Organon* in 1620:

> First then, away with antiquities, and citations or testimonies of authors; also with disputes and controversies and differing opinions; everything in short which is philological. Never cite an author except in a matter of doubtful credit: never introduce a controversy unless in a matter of great moment. And for all that concerns ornaments of speech, similitudes, treasury of eloquence, and such like emptinesses, let it be utterly dismissed. Also let all those things which are admitted be themselves set down briefly and concisely, so that they may be nothing less than words. For no man who is collecting and storing up materials for ship-building or the like, thinks of arranging them elegantly, as in a shop, and displaying them so as to please the eye; all his care

> is that they be sound and good, and that they be so arranged as to take up as little room as possible in the warehouse. And this is exactly what should be done here. (4.254–55)

Bacon makes supportive comments along similar lines in the *Advancement*.[70] Why did Bacon recommend plain style in communicating natural knowledge? One clue is given in the ship-building metaphor: this style is for use, not for show. A little later, in the *Preparative,* he returns to this theme: "In the history which I require and design, special care is to be taken that it be of wide range and made to the measure of the universe. For the world is not to be narrowed till it will go into the understanding (which has been done hitherto), but the understanding to be expanded and opened till it can take in the image of the world, as it is in fact" (4.255–56). Then, he compares natural historians to secretaries of canon law, "For we (who as faithful secretaries do but enter and set down the laws themselves of nature and nothing else) are content with brevity and almost compelled to it by the condition of things" (4.262). The natural historian channels "the image of the world": he is "compelled" to set it down faithfully, just as the Pythia was "compelled" to speak Apollo's words.

Considering Bacon's style recommendations in the light of prophetic ethos helps reconcile them. Prophets are compelled by the superhuman forces they channel not to elaborate their messages (plain style); however, when prophets teach the uninitiated, they must often cast hard-to-accept truths in terms of the familiar (fables and aphorisms). According to Perkins's method, Protestant preachers were likewise to employ these two modes when they (a) gave the "one entire and naturall sense" of the scriptures on the one hand and (b) induced from these scriptures principles for moral application using "allegories" on the other.[71] This bimodal prophetic discourse would not have seemed incoherent to Bacon's peers; it only seems so to scholars studying his rhetoric hundreds of years later.

Conclusion: Bacon's Prophetic Legacy

Although Bacon failed to gain adherents to his program of philosophical reform during his lifetime, he did succeed in alchemizing a new ethos—that of the experimental philosopher. He argued this ethos into form from rhetorical resources at hand—principally the ethos of the natural magician and that of the Protestant prophet. We saw that Protestant prophecy expressed the meme of classical prophecy with a few key mutations: particularly, an emphasis on textual sources and corporate validation. Bacon transferred the focus of these functions from the Book of God's Will to the Book of God's Works. So doing, he authorized "priests of Nature" to work in parallel with their religious brethren to restore English society. In his own life and writings, Bacon performed confirming signs of this ethos, which, like the luxurious mantle worn by the Father of Salomon's House, was ready to be taken up by his prophetic "sons" when the time was ripe. That auspicious kairos would open 50 years later, as the founders of the Royal Society sought to establish experimental philosophy as England's new civic oracle.

4 CONFIRMING SIGNS

THE PROPHETIC ETHOS OF THE EARLY ROYAL SOCIETY

In the famous frontispiece to Thomas Sprat's history (1667) of the fledgling Royal Society of London for the Improvement of Natural Knowledge—the world's longest-lived official scientific society—Francis Bacon is seated in the foreground with the motto "Renewer of the Arts" etched at his feet.[1] Decked out in his chancellor's insignia, he is positioned so that a bust of King Charles II, the Society's patron, and Viscount Brouncker, its first president, sit slightly behind him at his right hand. Bacon catches the viewer's eye and directs it to an array of symbols of operative knowledge taken from the *Advancement*: gunpowder for national defense, silk-making equipment for industry, and the compass for navigation and trade (figure 2).

The image makes a strong argument for Bacon's influence on the constitution of the Royal Society's ethos. It is an argument corroborated a few pages into the *History* by Abraham Cowley's dedicatory ode, which casts Bacon in the role of the Society's Moses. This argument is further reinforced by Sprat's text, which presents the Society as faithfully following Bacon's philosophical design and experimental method. Sprat's arguments turn out to reflect politics more than history[2]: the Society never seriously pursued Bacon's method.[3] And, although the "excellent Verulam" had promised funds in his will for the founding of a college that would have suited the Society's interests perfectly in its early years, Bacon's debts voided the trust.[4]

These facts and others have led twentieth-century science scholars to discount the Baconianism of the early Royal Society. As far as the Society's social organization and material practices go, they are clearly right. But there remains Bacon staring out at us from the frontispiece—as well as Cowley's ode, and the founders' protestations that they were Bacon's true prophetic heirs. I believe this rhetorical legacy deserves investigation in its own right, not just as an index to history. In addition to addressing the Society's early

FIGURE 2 The frontispiece to Thomas Sprat's *History of the Royal Society* (1667), engraved by Wenceslas Hollar from a design most likely by John Evelyn. Francis Bacon is seated at the right with the title "Renewer of the Arts." William Brouncker, Viscount, is seated on the left with the title "Society President." The bust between them is of King Charles II, who is given the title "Society Patron." Bacon points with his left hand to an array of inventions he had touted in his *Advancement of Learning* (1605) as validating the utility of the new experimental science: guns and gunpowder, silk-making equipment, and the compass. The shelves of books on the left of the image, and Boyle's first air pump on the ledge behind them, symbolize the Society's dual Baconian commitment to divining natural truth from reliable natural histories and from experimentation.

Source: The Royal Society, © The Royal Society.

Baconian rhetoric, I engage a few additional outstanding questions about it, including the source of its ethos, the role of women in the constitution of that ethos, and how it was performed in the first volume of the Society's journal, the *Philosophical Transactions of the Royal Society* (*PTRS*).

In the remainder of this chapter, I argue that Bacon was more than just an authorizing figurehead for the Society founders. They imitated many of the prophetic strategies he had recommended and performed. They did this not just because Bacon had told them to but because they witnessed first-hand the power of prophetic ethos: in a tenuous political climate, it enabled the Society to transcend a host of political dilemmas—framed by antinomies such as public/private, experience/authority, and contemplation/action—to establish itself as a civic oracle in Restoration London.[5] The founders' successful hybridization of prophetic and experimental ethos still conditions the ethos of our science advisers today, as we will see in ensuing chapters.

Ascertainment: Converting Uncertainty to Conviction

The 1660s were a tumultuous time in London, despite the initial hopes for peace that came with the Restoration. The two most famous disasters were the Plague of 1665 and the Great Fire of London in 1666, but the Second Anglo-Dutch War (1665–1667) was also a stressor. The religious landscape was no more serene: the backlash against Protestant reformers following Charles II's Restoration spurred religious emigrations to America whose numbers swelled with droves of Scots escaping famine up north.[6] In short, these were uncertain times, and people were desperate for certain knowledge of goings-on and future prospects.

As in all crises, prophetic figures emerged to meet the demand for certainty, declaiming either jeremiads or isaiads depending on their creed.[7] These figures addressed moral, social, and political questions, but what about the natural disasters? These were queries for prophets of natural philosophy. But the philosophers themselves disagreed about what counted as certainty in natural knowledge and what was therefore the proper end of natural philosophy. Thomas Hobbes, Margaret Cavendish, Meric Causaubon, and others argued that natural philosophy should show how natural phenomena were generated by a priori "final causes." These were revealed in the Bible and in the works of Aristotle (who was ironically now received as a Christian authority after having taken several centuries' worth of abuse from the Church). Such a method generated certainty by showing how an event such as an earthquake fit into God's unchanging plan for the universe. By rejecting this dogmatism, a natural philosopher risked being branded an atheist, a position that would not be fashionable—or safe—to hold for another hundred years at the very least. And indeed, skeptics following in the tradition of Sextus Empiricus or Michel de Montaigne did sometimes go so far as to argue that nothing could be true unless it were directly sensed, an argument that created problems for belief in God or magnetism. So, skepticism was not a viable alternative to dogmatism for the Christian gentlemen who wished to found an experimental society.

The founders transcended the dogmatism/skepticism antinomy by performing prophetic Ascertainment. They refused to hew to Aristotelian explanations of natural causes and effects. But in rejecting Aristotle, they were not rejecting the possibility of reliable natural knowledge. Instead, they gathered reasonable men together and performed experiments whose witnessing grounded reliable belief in the causes and effects of natural phenomena. This process did not yield absolute knowledge of nature. Instead, it instilled "moral certainty," a conviction in things seen, both confident and humble, definite yet admitting of correction and revision; such certainty was sufficient to found actions that could benefit the commonwealth.

As Henry Van Leeuwen has demonstrated, the Society's solution to the dogmatism/skepticism dilemma was outlined in two tracts, John Wilkins's *Of the Principles and Duties of Natural Religion* (published posthumously in 1675) and Joseph Glanvill's *Scepsis Scientifica* (1664).[8] Wilkins was so important to the founding of the Royal Society that some scholars have opined it could not have happened without him. He convened the Oxford Circle and first invited Robert Boyle to it, and he was so dedicated to the cause of founding Bacon's Salomon's House that he continued to press its case with the very king who had removed him from his presidency of Trinity College.

Principles and Duties was aimed less at the Aristotelian dogmatists than at "that Humour of Scepticism and Infidelity, which hath of late so much abounded in the world, not only amongst sensual men of the vulgar sort, but even amongst those who pretend to a more than ordinary measure of wit and learning."[9] The antidote to this destructive skepticism was natural religion, or the general revelation that God had made in the physical world of his Providence for humanity. Wilkins admitted this revelation could not be induced by men infallibly; nevertheless, through the evidence of logic, experiment, and testimony from trusted witnesses, philosophers could attain "indubitable certainty" about Nature's order, and this certainty, which he elsewhere called "moral certainty," would lead them to worship God and undertake works to better Creation.

Glanvill was a young loyalist whose long-standing friendships with Wilkins and other Puritan founders testify to the Society's commitment to toleration and solidarity. *Scepsis Scientifica* was an update to Glanvill's *The Vanity of Dogmatizing* (1661). In the reissue, Glanvill professed himself astonished that he had been branded a skeptic by Catholic inquisitors as a result of his criticism of Aristotle's physics. Glanvill retorted "that there's a *medium* between being *blind* and *infallible*"[10] and pointed to the Royal Society's experimental methods as an illustration of the ideal mediating epistemology. These methods he treated in more detail in *Plus Ultra* (1668), presenting them, as Bacon did, as a philosophic analog to current prophetic practice: fundamental devotion to the evidence of the experiment, cautious and communal induction of causes and effects, and application of these principles to the betterment of the commonwealth. Glanvill swore of the Society fellows that "their Designs and Methods of Inquiry naturally form men into the modest temper, and secure them from the danger of the quarrelsome Genius." In other words, the Society's methods marked out a middle path that steered it clear of the Charybdis of dogmatism while shielding it from the snapping Scyllic heads of atheism and enthusiasm.[11]

These were primarily arguments for the way the Society pursued certainty internally, but they also applied to petitions for certainty it received from powerful factions in London. Rhetorically speaking, what the experimental operations did was transform requests for certain knowledge into a demonstration of the value of the experimental ethos. For example, King Charles II, the Society's patron and most regular petitioner, frequently sent items from his curiosity cabinet to have their purported powers ascertained. One such item was a poisoned dagger, which the Society tested by cutting a kitten with its tip; the kitten fell asleep but then revived.[12] Prince Rupert likewise sent some of his inventions to be tested, including a new formula for gunpowder and a "water-raising engine."[13] These tests were witnessed not only by the Society fellows but occasionally by the royal petitioners themselves and state visitors who were brought to Trinity House (after the Fire, Arundell House) as part of the grand London tour. Through this process of communal witness and validation, the experiments established consensual "matters of fact";[14] these settled some questions about nature while moving the drama on to new ones: for example, could the explosive capacity of Rupert's gunpowder be improved? Would *more* poison kill the kitten? The dialogues opened in this way led to future experiments, which spurred future dialogues, and so on.[15] Through the persistence of this interaction, the royals performed moral certainty about the matters of fact demonstrated by the Society; ergo, they performed moral certainty about the Society's value to England.[16]

The members of the royal family were not the only agents professing their faith in this way. The government and military frequently consulted the Society for more certain methods and machines. In the words of Steven Shapin and Simon Schaffer:

> The wants addressed by the experimental community spread across Restoration economic, political, religious, and cultural activity. Did gunners want their artillery pieces to fire more accurately? They then should bring their practical problems to the physicists of the Royal Society. Did brewers want a more reliable ale? Then they should come to the chemists. Did physicians want a theoretical framework for the

explanation and treatment of fever? Then they should inspect the wares of the mechanical philosopher.[17]

Again, what the Society provided in any of these instances was not certainty in the sense of absolute knowledge. While their early work did indeed yield a few useful explanations of phenomena and inventions (for instance, Boyle's air pump and the resulting law relating pressure and temperature), most of the experiments published in the *PTRS* were too haphazardly performed or too colored by folk wisdom to produce useful insights.[18] But through its experimental demonstrations and written accounts, the Society did provide a rhetorical occasion for powerful factions in London to develop and demonstrate their faith in the new experimental philosophy. The case of the "sea watch" illustrates this process nicely.

One of the most intense "wants" in English society was improved navigation. The burgeoning merchant marine wanted more certain ways of measuring longitude to guide their transoceanic voyages; more accurate navigation meant more trips and thus more money. Mariners came to Society fellows for help devising a longitude chronometer: Robert Hooke was one of the first founders to respond to these requests, with a balance-spring design that obviated the problems that pendulums encountered on the rolling sea. However, he lost a patent fight over the work and abandoned the project in disgust. But petitions were still coming in, and so other fellows took up the problem, including Christiaan Huygens and Edmund Halley. As Dava Sobel details in her recent history of this drama, balance-spring designs were tricky affairs, and so the development of a working chronometer was a drawn-out process that spilled outside the bounds of the Society. It became spectacularly politicized when Parliament passed the Longitude Act in 1714 and offered up to £30,000 for a reliable sea watch. After a series of dramatic negotiations with the Board of Longitude and battles with rivals, in the 1760s the clockmaker John Harrison finally won a partial award—less than he deserved—for the first chronometer to reliably calculate longitude.[19]

The drama of the chronometer shows us how the Society performed Ascertainment for England. Powerful factions in the country were already interested in extending human control over nature through experimentation and invention. With their petitions to the Society, they initiated a dialogic process that performed their support of the new experimental philosophy. English merchants thus obeyed the old Delphic motto "know thyself" by broadcasting their political convictions. The Society could only facilitate this process by maintaining a privileged authority independent of the fray of politics as usual[20]—much like Bacon's Salomon's House.

Authorization: Salomon's House Again

In 1641 the famed Puritan educational reformer Jan Comenius beseeched his friend Samuel Hartlib to press the current king to institute Bacon's reforms: "I beg you, reread the introduction to the second book of his *Advancement of Learning*; and see whether the entreaties there to King James . . . may not with better authority be conveyed to the most serene Charles . . . , particularly when things are to such an extent changing for the better, as is the hope of all good men."[21] Alas, Comenius did not seem to share Bacon's prophetic gift, because within a year England was once again plunged into civil war. The privilege of seeing Bacon's philosophical reforms become reality would fall to the next generation of experimental philosophers, who received the king's charter for the Royal Society in 1662, the year Hartlib died.

Hartlib and his colleagues were Baconians of the first order.[22] Though the generation immediately succeeding Bacon had not paid his arguments for the reform of natural philosophy much mind, the Interregnum resuscitated the synergy between Protestant millennialist reform and operative natural philosophy—this time guided by ascetic Puritans.[23] Reformers again dreamed of an instauration of English society, as Milton demonstrated in his famous *Areopagitica* speech to Parliament in 1644:

> Behold now this vast City; a City of refuge, the mansion house of liberty, encompast and surrounded with his protection; the shop of warre hath not there more anvils and hammers waking, to fashion out the plates and instruments of armed Justice in defence of beleaguer'd Truth, then there be pens and heads there, sitting by their studious lamps, musing, searching, revolving new nations and idea's. . . . What could a man require more from a Nation so pliant and so prone to seek after knowledge. What wants there to such a towardly and pregnant soile, but wise and faithfull labourers, to make a knowing people, a Nation of Prophets, of Sages, and of Worthies. We reck'n more than five months yet to harvest; there need not be five weeks, had we but eyes to lift up, the fields are white already.[24]

During the short-lived, harsh social stability enforced by Oliver Cromwell's regime, two interwoven groups of powerful gentlemen-scholars "seek[ing] after knowledge" coalesced—one around Hartlib and one at Oxford's Gresham College—to pursue experimental programs that could usher in the New Jerusalem. Given these goals, it is not surprising that both groups circulated calls to influential people in Cromwell's regime for the founding of Bacon's Salomon's House in London.[25] The prophetic motif seemed to resonate with the Lord Protector: Hartlib reported a favorable response to his proposal to buy land for a "College of the Six Days' Work," and Parliament authorized the formation of a Committee for the Advancement of Learning in 1653, which Hartlib's colleague John Dury had recommended "to put in practise the Lord Verulam's Designations *De Augmentis Scientiarium* among the learned."[26]

The Protectorate, too, fell apart, but by the time it did, the movement to found a state-supported institute of experimental philosophy had picked up enough momentum for its adherents to press their case with Charles II and the agents of the Restoration. As soon as the new king ascended to the throne in 1660, the fellows of the Hartlib and Oxford circles began working in earnest to get a royal charter for their "Salomon's House." We have surviving drafts of the crest, motto, and charter that were to be presented for Charles's approval, and the prophetic references in them to Bacon are striking. An early draft of the motto was *et augebitur scientia* (and knowledge increases), a direct allusion to the frontispiece of the *New Organon*, on which Bacon had had inscribed Daniel 12:4.[27] This prophecy read *in toto*: "In those days many shall run to and fro, and knowledge will increase." Protestant reformers had interpreted this prophecy since the Late Tudor period to refer to the boom in sea trade, which they believed augured a period of intense scientific reform, which would in turn yield the New Jerusalem. Christopher Wren's draft of the charter for the Royal Society likewise borrowed heavily from prophetic ethos and from Bacon:

> The Way to so happy a Government, we are sensible is in no Manner more facilitated than by the promoting of useful Arts and Sciences, which, upon mature Inspection, are found to be the Basis of civil Communities, and free Governments, and which gather Multitudes, by an Orphean Charm, into cities, and connect them in Companies; that so, by laying in a Stock, as it were, of several Arts, and Methods of Industry, the whole Body may be supplied by a mutual Commerce of each other's peculiar Faculties; and consequently that the various Miseries, and Toils of this frail Life, may, by as many various Expedients, ready at Hand, be remedied, or alleviated; and Wealth and Plenty diffused in just Proportion to every one's Industry, that is, to every one's Deserts.[28]

The charter goes on to argue that such a project will put England ahead of "barbarous nations" that are otherwise equals. It finishes,

> We purpose to make further provision for this branch of knowledge likewise, Natural Experimental Philosophy; which comprehends all that is required towards those intentions we have recited; taking care in the first place for religion, so next, for the riches and ornament of our kingdoms.[29]

The operative and millennial ends of Baconian experimentation couldn't be more clearly articulated here.[30]

In this crucial public performance, and elsewhere, the founders presented themselves as taking up Bacon's prophetic mantle.[31] Were the founders seeking simply to bask in Bacon's revived glory? Too much was at stake during these turbulent times. Rather, it makes more sense to hypothesize that the founders realized that performing prophetic ethos, as Bacon had, could secure a place for their new Salomon's House in the midst of a widespread reevaluation of traditional sources of authority—religious, social, political, and philsophical.[32] A particular dilemma the Royal Society founders faced in the arena of authority focused on the proper role of the Christian gentleman. Many factors contributed to this dilemma, but major (and related) ones were the rise of the middle class and Reformation arguments about the relative merits of work and leisure.

The middle class had been growing since Henry VIII's conciliatory reforms, which constituted the House of Commons and extended (slightly) suffrage. Increasing trade had also put more wealth at the disposal of the merchant class, who bought university educations for their sons and thus blurred the distinction between the bourgeois and gentry. This development put landed gentry in England at pains to demonstrate why they merited the honor and rents that historically had been "due" them as their birthright.[33] Grubbing around in laboratories or in the woods collecting samples hardly constituted a performance of gentility. We plainly see anxiety about the "baseness" of these activities reflected in first Bacon's and then Boyle's extensive attempts to defend the consonance of experimental philosophy and gentle status.[34]

Meanwhile, the English Reformation, inflected by Calvinist theology, had made industry rather than leisure the mark of the true Christian.[35] Under this paradigm, Christian gentlemen were under pressure to prove they had been elected to salvation by performing good works that bore fruit for the commonwealth. Thus the universities, long suspected of being a haven for Papists, were now doubly suspicious because the philosophers there lived the classical *vita contemplativa*, or contemplative life, rather than the *vita activa* of political engagement. In his Easter recess speech to Parliament in 1607, James I sniped at university philosophers for putting their professional efforts to "trial[s] of [their] wit" rather than public welfare.[36]

This was the contemplation/action antinomy against which the Society founders struggled to situate their new enterprise.[37] If the Society were isolated from active life, as Oxbridge was, it would sever its lifeline to the Protestant reformers, who had powerfully revived Bacon's program. However, if the Society took to "fiddling with air-pumps," as Thomas Hobbes disdainfully characterized experimental labor,[38] it risked demeaning itself under the old paradigm of gentle ethos. While they were happy to employ the educated and active community of London tradesmen—who were at that time making stunning advances in lens grinding, clock making, dyestuffs, and other industrial applications—the gentlemen who founded the Society were not at all keen to be classed with them.[39]

The founders transcended this ethical dilemma by cultivating prophetic ethos, specifically by performing the role Bacon had recommended, the "priest of Nature" (14.26). Just as Protestant priests served their parishes while being provided with a living so they could contemplate the divine Word, so the founders imagined being supported while they contemplated the World.[40]

The founders applied this argument not only to securing a space for Salomon's House but also to individual ethos, as they attempted to persuade other Christian gentlemen to join their project. For this recruitment effort, they relied on a pair of arguments popularized by Bacon's sermon on Ecclesiastes 3:11 in the *Advancement* (which I discussed in chapter 3): first, that God had hidden evidence of his Providence as "hieroglyphicks" in nature, which it was the English gentlemen's Christian duty to decode; second, that this decoding of God's Providence would automatically yield two kinds of fruit—"wonders" that would compel belief from nonbelievers and "goods" that would benefit the polity.[41]

This argument couplet became the rhetorical signature of the early Society, repeated in the published tracts of Robert Boyle, John Wilkins, Joseph Glanvill, Walter Charleton, and others.[42] It transcended the contemplation/action antinomy by showing that good works organically followed from contemplation of the Book of Nature. The Christian gentleman who inhabited this ethos was the true priest of Nature. As Robert Boyle derived this ethos, "It will not be rash to infer that if the world be a Temple, man sure must be the Priest, ordeain'd (by being qualified) to celebr[a]te Divine service not only in it, but for it."[43] The Priest of Nature had to stay out of politics in order to dedicate himself to this "Divine service," but ironically, that position enabled him better to profit the polity by the good works that flowed unimpeded through him from God's Providence.[44]

In *A Social History of Truth*, Steven Shapin also notes the power of the "priest of Nature" ethos, but overall, he ascribes the success of the Society's ethos to its identification with elements of gentle ethos. This ethos provided many of the same benefits as prophetic ethos: it served as a guarantee of both trustworthiness and special access to natural knowledge, conflated contemplation and action, and secured its performers with a stable and privileged position in seventeenth-century England.[45] But as should be apparent from Boyle's equation of the Christian gentleman with a "priest of Nature," prophetic and gentle ethos were not exclusive at this time. So, my analysis runs in parallel with Shapin's to this point. Where our explanations start to generate different predictions, however, is when they encounter agents involved in the constitution of experimental ethos who were not, strictly speaking, Christian gentlemen—namely, technicians and women.

In the case of technicians who could not perform gentle ethos, Shapin alots them small share in the experimental ethos, arguing that their substantial contribution to the Society's work was generally credited to the gentlemen experimenters that employed them under the provisos of the old master/servant ethical tradition.[46] By contrast, my hypothesis would predict that anyone who could demonstrate privileged access to natural knowledge, and could leverage this knowledge to engage the polity in dialogue, would be recognized as an experimental philosopher. The case of Robert Hooke is telling in this regard.

True, compared to many of the titled Society founders, Hooke was born in humble circumstances, the son of a clergyman on the Isle of Wight. However, his extraordinary intelligence won him a position at Oxford and the notice of Robert Boyle, who employed him as a technician. (Hooke was the only person, in fact, who could reliably get Boyle's air pump to work.)[47] When the Society was chartered, Hooke was named Curator of Experiments. He invented the balance-spring design for chronometers, as mentioned above, and he published an astonishing and popular collection of etchings made from his microscopic observations, the *Micrographia* (1667), a copy of which was gifted to Charles II and which Samuel Pepys declared the "most Ingenious book I ever read in my life."[48] Hooke was granted the Cutlerian lecture in mechanics at Oxford, and became Gresham Professor of Geometry there and, eventually, Doctor of Physic in 1691. He died in 1703 a wealthy man whose social status was nearly equal to that of his titled fellows. It is perhaps an indicator of Hooke's power in the Society that Isaac Newton, with whom Hooke had a long-standing feud, did not ascend to the presidency until after Hooke's death. In retribution, Newton immediately removed Hooke's experimental folio from the Society library, to which it was only recently returned after its discovery in a Hampshire attic in 2006.

This history constitutes more than a mere "technician." Perhaps it is because Shapin wrote his book before the rediscovery of the Hooke folio that he downplayed Hooke's ethos in the Society. At any rate, it is now clear from the historical record that Hooke played a key role in the formation of the Society's early ethos and practices.[49] And if Hooke did not perform gentle ethos, he certainly performed prophetic ethos, as we will see when we discuss the *Micrographia* in more detail later, and as we can see in this argument from its preface:

> By the addition of such artificial Instruments and methods, there may be, in some manner, reparation made for the mischiefs, and imperfection, mankind has drawn upon it self. . . . The only way which now remains for us to recover some degree of those former perfections, seems to be, by rectifying the operations of the Sense, the Memory, and Reason, since upon the evidence, the strength, the integrity, and the right correspondence of all these, all the light, by which our actions are to be guided, is to be renewed, and all our command over things is to be establisht.[50]

It would be hard to find a clearer example of the classical isaiad: the indictment of humanity's "mischiefs," followed by a promise of redemption if philosophers will but turn from their sinful disregard of the evidence of the senses and set themselves to reading God's natural revelations.

Through these arguments and the visual evidence of his extraordinary mechanical and observational abilities in the *Micrographia* (figure 3), Hooke was able to engage both his Society fellows and powerful outsiders like Pepys in a dialogue that validated his special access to natural knowledge and his ethos as an experimenter. Hooke's case reminds us of the importance of considering as many factors as we can in the constitution of ethos. True, the Society's official experimental reports do not figure Hooke, Dennis Papin, or other key technicians as owning experimental ethos in any significant way. However, if we consider texts from the wider kairos—Society council minutes, patent drafts, folios such as Hooke's, and writings about the Society by outsiders—we can appreciate the agency of technicians in constituting experimental ethos, even if, as Deborah Harkness suggests, they served as "foils" for the founders' arguments about the ideal form of this ethos.[51]

Similarly, gentle ethos as Shapin constructs it—partaking of the old Germanic warrior's code and Calvinist predestination—would predict no place for women in administering the experimental ethos. He argues that while gentlewomen were accorded many of the same benefits as their male cohorts—including private tutors at least through the grammar school years—they had no official voice in any polity governing England, including the

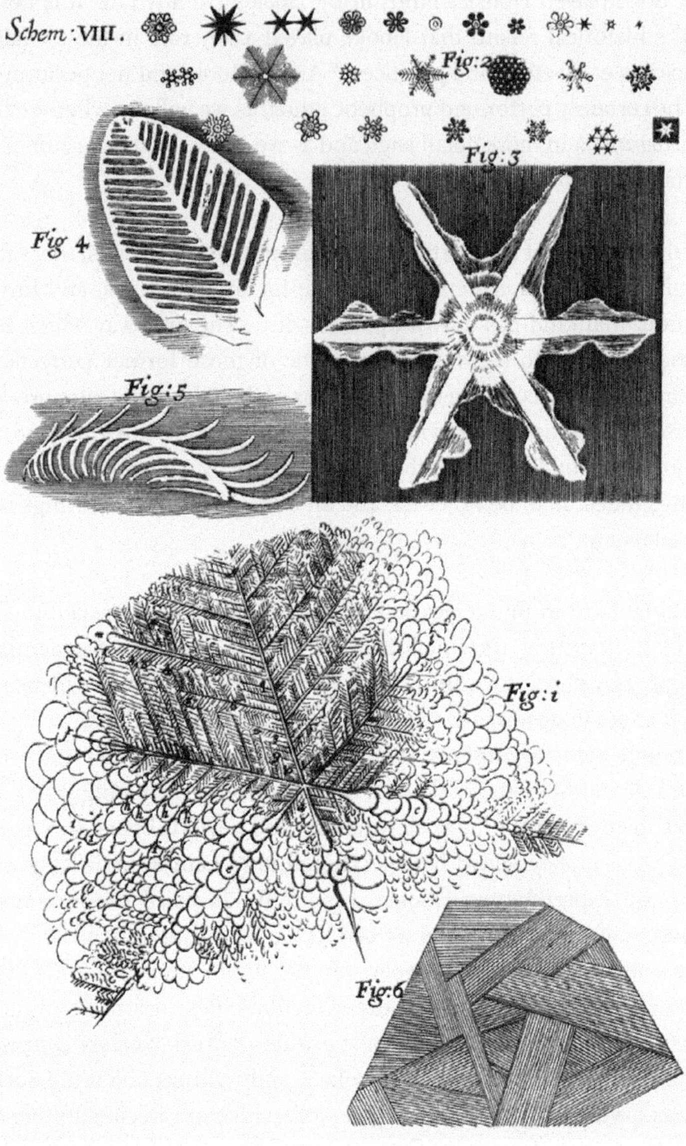

FIGURE 3 Scheme 8 from Robert Hooke's *Micrographia* (1665), engraved from his hand drawings of microscopic views of frozen water (snowflakes, frost structures, etc.).

Royal Society, in which women were relegated to "invisible" supporting roles. Shapin dismisses as Boyle's "greatest technician" his sister Lady Ranelagh, with whom Boyle lived for the majority of his tenure in the Society, and who hosted his colleagues in her home.[52] And indeed, the Society's Charter Book does not show a woman's signature until 1945. However, if we draw back from a narrow view of the founders' writings, we discover that women had a much more active role in shaping the Society's prophetic ethos.

David Noble admits that the Society's "celibate, homosocial, and misogynous" culture excluded women from its official activities, but he has uncovered influential women in the surrounding kairos.[53] As we briefly discussed, the Protestant Reformation loosened strictures on women's role in public religious discourse. During the Interregnum, when the Gresham and Oxford circles were forming, women directed matriarchal prophetic sects, such as the Quakers, and they published tracts on natural magic and healing, such as Elisabeth Grey's, which went through more editions in the seventeenth century than Bacon's famous *New Atlantis*. These women were generally part of the artisan classes, but aristocratic women also visibly supported science by holding salons in their houses, as Lady Ranelagh did. Nor did they restrict themselves to a supporting role. Shapin does not mention this, but Boyle's sisters Lady Ranelagh and Mary Rich authored a medical treatise, complete with an herbal guide and alchemical notations.[54]

Another of these important scientific texts was Margaret Cavendish, the Duchess of Newcastle's, *Observations upon Experimental Philosophy* (1666).[55] Cavendish bears the distinction of being the first woman officially invited into Society chambers, in May 1667. Samuel Pepys reported the visit as an uncomfortable one for everyone, griping "nor did I hear her say any thing that was worth hearing, but that she was full of admiration, all admiration."[56] And yet there is more to Cavendish's case than meets the eye, as there is with the other women who drifted like "ghosts" through the written records of the early Society.[57] Well-educated in a family famous for its scholarship in natural philosophy, Cavendish followed scientific theories and developments closely, though she came to a different conclusion than the Society did about the utility of the new experimental method. In *Observations* she sided with her tutor Thomas Hobbes in criticizing Boyle's theories about vacuums as violating Aristotelian plenism (the argument that "Nature abhorred a vacuum," so seemingly empty spaces were in fact filled with ether). In general, she was dubious that experiments could yield truth. In an argument that foreshadowed constructivist critiques of experimental science by 300 years, Cavendish doubted the images published in Hooke's *Micrographia* because she believed the lenses distorted reality; she argued, "Art doth more easily alter than inform."[58] She also railed against the mistreatment of animals, which the Society was clearly guilty of in its early years as it performed inadequately anesthetized vivisections of dogs and other gruesome natural auguries.[59]

Her public arguments against this "cruelty" were echoed in her *Blazing World* (1668) and *World's Olio* (1655), both of which had a decidedly prophetic cant. *Blazing World* was addressed to women as a commentary on *Observations*, designed to be joined to it as "Two several Worlds, at their Two Poles."[60] In contrast to the masculine experimental paradigm depicted in *Observations*, *Blazing World* is Cavendish's feminist Bensalem. It records her ecstatic vision of a matriarchal world whose Empress allows her subjects to pursue natural philosophy for their delight, but not as a means to truth, which can only be attained through charity and faith. To these virtues the Empress coaxes her people using miraculous demonstrations of natural magic:

> And thus the Empress, by Art, and her own Ingenuity, did not onely convert the Blazing-World to her own Religion, but kept them in a constant belief, without inforcement or blood-shed; for she knew well, that belief was a thing not to be forced or pressed upon the people, but to be instilled into their minds by gentle perswasions; and after this manner she encouraged them also in all other duties and employments: for Fear, though it makes people obey, yet does it not last so long, nor is it so sure a means to keep them to their duties, as Love.[61]

She had first harped on the contrast between the angelic "Love, Piety, Charity, Clemency, Patience, Humility" of women's persuasive ways versus the devilish "Ambitions, Extortion, Fury, and Cruelty" of men's government in *World's Olio*.[62] In the bizarrely coy introduction to this document, Cavendish inhabits by turns two ethical stances: one that demurely dramatizes men's superiority, employing all the usual antinomies of hot/cold, hard/soft, oak/willow, and so on, and one that launches a devastating attack against her sex's "enslavement" by men and relishes the notion of their government being brought down by the million stings of women's "effeminate monarchy."[63] And this is just the introduction. The *Olio* proper is a collection of aphorisms in the Baconian tradition on topics such as "Fame," "Wit," "Fancy," and "Valour"—in other words, wisdom literature. This was a bold statement indeed, to write wisdom literature after a lengthy preamble declaring women incapable of doing so.

Although we have no evidence that the Society took Cavendish's critiques seriously,[64] hers was not the only woman's voice echoing through the halls of Arundell house, calling for an alternate, inclusive, and compassionate science. Anna Barbauld, one of Joseph Priestley's assistants in the experiments leading to the discovery of oxygen, wrote a poem from the point of view of one of the mice Priestley sacrificed for his oxygen trials and stuck it on the bars of its cage for Priestley to see. The last stanza read, "The well-taught philosophic mind / To all Compassion gives; / Casts round the world an Equal eye, / And feels for all that lives."[65] This poem is just one of many documents that historian Richard Holmes has recently gleaned from the Society's archives that reveal women interceding on behalf of marginalized Others, both human and non-.[66]

It seems quite clear that the "clerical" cultic establishment represented by the Society fellows found this discourse threatening. Beginning with Francis Bacon, the Society had insisted on feminizing Nature and describing its work on her in terms of insemination and rape.[67] This narrative paralleled a troubling backlash against the power that women prophets and healers had gained during the Reformation; the Counter-Reformation rebranded them "witches" and killed over a million of them, by some estimates, across Europe and America.[68] Sadly, some Society fellows, particularly Robert Boyle and Joseph Glanvill, lent indirect support to this fatal misogyny by arguing for the threatening reality of witchcraft.[69]

But if the Society suppressed women's voices, it could not silence them, just as it could not divorce its ethos from that of the technicians whose work it relied on. In this way both women and technicians helped to constitute the experimental ethos in the seventeenth century by maintaining a kairotic counternarrative of the feminine, the common, and the silenced against the Society's cultic narrative of the masculine, elite, and official. We will see this counternarrative amplified to a nation-shaking pitch in Rachel Carson's protests against indiscriminate pesticide use in chapter 7.

This was the opportunity cost for the founders of adopting prophetic ethos: if it offered them a way to transcend many of the dilemmas they faced, it also made it impossible to maintain a monolithic, authoritarian discourse of the masculine and privileged. But the Otherness that echoed in its voice was itself a confirming sign of the Society's extraordinary social authority, just as it had been for the Delphic oracle.

Confirmation and Divination: The Air Pump and the Microscope

In 1659 Sir Robert Boyle demonstrated his new air pump before a group of gentlemen who would soon be Society fellows. As they watched, Boyle's assistant Robert Hooke lowered a songbird into the pump's glass receiver:

> The vessel being hastily, but carefully clos'd, the pump was diligently ply'd, and the bird for a while appear'd lively enough; but upon a greater exsuction of the air, she began manifestly to droop and appear sick, and very soon after was taken with . . . violent and irregular convulsions . . . [S]he seem'd to be dead within seven minutes . . . but upon the

> speedy turning of the key, the fresh air flowing in began slowly to revive her, so that, after some pantings she open'd her eyes, and regain'd her feet, and in about a 1/4 of an hour after, threatened to make an escape at the top of the glass.[70]

This miraculous performance was repeated multiple times for fellows of the Society and distinguished visitors.[71] Because the early Royal Society relied so heavily on operative performances of specialized equipment like the air pump to confirm their privileged access to natural knowledge, the topoi of Confirmation and Divination reveal largely the same vista when applied to the Society's rhetorical practices—namely, strategies mounted in reaction to conflicting arguments about the proper source of reliable natural knowledge.

As discussed above with respect to Ascertainment, seventeenth-century England was a hothouse of epistemological debate. On the one hand, the Church had redeemed the peripatetic philosophers as sanctioned authorities, and so Aristotle's words on physics and Hippocrates's on medicine now stood as the last word. They articulated "final causes" of nature that could not change because God did not change. On the other hand, Galileo's telescopic experiments and the works of the natural magicians were circulating widely, and the wonders these men had seen with their own eyes and made with their own hands were as compelling as they were at odds with Aristotle.[72]

The Society founders struggled to reassure the Church that their activities were not heretical as they struggled not to fall behind the experimental zeitgeist building around them in France, Italy, and Germany. The motto the Society finally chose was *Nullius in verba*, implying that the philosophers would take no claim about natural law on previous authority; rather, they would test it before they gave it credence. This was a frequent caution in Bacon's works, but it did not sit well with some of their peers in the clergy, including Meric Casaubon, and with continental critics such as Benedict de Spinoza.[73] This was the ethical dilemma framed by the experience/authority and heresy/piety antinomies.

The Society employed many tactics in confronting this dilemma, but the most effective were confirming signs because in them phenomenology mooted epistemology: that is, the natural wonders the Society produced as "matters of fact" made it hard for critics to argue that they were not using the right methods.

These signs were achieved via a new divinatory method—experiments that employed instruments to remove individual bias as an obstacle to seeing natural order clearly. Since the Church had condemned the enthusiasm of natural magic, Society founders touted the rigorous ethical discipline imposed by "helps," in Bacon's terms,[74] such as telescopes, microscopes, air pumps, and exacting procedures. As we saw earlier, Hooke stressed in the preface to *Micrographia* that his microscope corrected sensory defects humans suffered by their sinful Fall from Eden.[75] Wren and Boyle made nearly identical arguments,[76] with Wren going so far as to argue that as opposed to the "vanity" of current doomsday prophecies circulating in London's streets, "experiment and reason [are] the only way of prophesying natural events."[77]

In a bizarre and wonderful passage from *Occasional Reflections*, Boyle explicated this divinatory technique by which readings from the book of Nature could be converted into spiritual lessons:

> 'Twas doubtless a very great pleasure to Aesop, that by his ingenious fictions he could, in a manner, lend reason and speech to lions, foxes, crows, and other animals, to whom Nature had deni'd both; and I know not why it should be less delightful, by Occasional Reflections, to turn not onely birds and beasts, but all kinds of creatures in the world, as well mute and inanimate, as irrational, not onely into teachers of ethicks, but oftentimes into doctors of divinity, and by compelling senseless creatures to reveal truths to us, that they were never acquainted with themselves, perform really something like that, which was but pretended by the antient augures and other diviners the heathen world admir'd, who took upon them, by the casual flights of birds, and the inspection of the intrals of beasts, to learn the will of Heaven. 'Tis . . . neither a less noble nor a less delightful piece of skill, to be able, by an innocent kind of necromancy, to consult the dead, and conjure up worm-eaten carkases out of their mossy graves, without fearing to hear from them such dismal Discourses as Saul had from dead Samuel, and to make, not the stars onely, but all the creatures of Nature, and the various ocurrences that can fall under our notice, conspire to inrich us with instructions they never meant us . . .[78]

Boyle here describes a kind of natural augury, a reading of the inner workings of natural objects to induce from them the grammar by which they were written. And indeed, the first volume of the *PTRS* is full of reports of dissections of "monstrous births," such as two-headed calves and other prodigies. Such natural divinations had yielded Boyle and other Protestant gentlemen private spiritual confirmation of God's Providence; now, by making them public, the Royal Society converted vivisections of dogs or suffocations of sparrows into confirmation of its prophetic vocation.[79]

There is no doubt among scholars who have studied these demonstrations that they were aimed at securing political status and support—and that they succeeded in this function.[80] These "natural wonders" included inducing luminescence in rotting fish, suffocating sparrows, making magnets cohere, mixing chemicals to change their colors, killing slugs by pouring salt on them, and digesting meat with acid. While many of these sound more like the pranks of 12-year-olds than scientific experiments, each represented a natural manipulation that had not been performed and observed systematically until this time. The more elaborate demonstrations, like the air pump, tended to be reserved for royals and distinguished visitors, but many demonstrations were performed in the general meetings, which anyone could attend.[81]

Even so, the audiences for the real-time demonstrations were limited, and the Society needed to expand its support base. Accordingly, the demonstrations were duplicated in print—in Hooke's *Micrographia*, Wilkins's *Real Character*, and the *PTRS*—via elaborate engravings. Steven Shapin and Simon Schaffer studied the persuasive effect of engravings of Boyle's air pump in the *PTRS* and concluded that these facilitated "virtual witnessing": that is, viewing the engravings served as a proxy for the actual experience of watching Hooke suffocate a sparrow in the pump's glass chamber.[82] Via the technology of print, the founders extended the polity that recognized their authority to include merchants, ambassadors, Continental philosophers, and American plantation owners, who might not be able to attend general meetings but could support the Society by subscribing to the *PTRS* or by financing experiments that might yield useful technologies for their trades.[83]

What was true of the air-pump engravings was even more so for the engravings in Hooke's *Micrographia*, in which he displayed intricate patterns too small for the unaided human eye to discern—the reticulated eye of a fly, the fingernail-rough scales of a fish, the crystalline architecture of frost (figure 3). Comparisons between these natural structures and equivalent human inventions—telescope lenses, armor, embroidery—were inevitable, and Hooke encouraged them in the text accompanying the images.[84] *Micrographia* made a strong argument for the Society's privileged ability to reveal God's Providence in the "secret workings" of nature.[85] Viewers critical of Hooke's methods were nevertheless hard pressed to deny the confirming power of his revelations, or to refute them without engaging the same techniques. That these natural wonders were taken as the cardinal sign of the Royal Society we know from how frequently they were satirized. For instance, the protagonist of Thomas Shadwell's satire *The Virtuoso* is named Sir Nicholas Gimcrack—a "gimcrack" being a false spectacle. And the poet Samuel Butler in his fragment "A Satire Upon the Royal Society" mocked the philosophers for spending their energies to discover "If chemists from a rose's ashes / Can raise the rose itself in glasses."[86] The learned in London, whether they approved of experimental philosophy or not, recognized the natural wonders the Society produced as the calling card of a new profession. The wide public recognition of the matters of fact created by the Society begged questions about the correct source of this knowledge: here was proof of success, in the suffocated sparrow or the flea's feathery armor.

A powerful supporting confirmation of the Society's legitimacy could be found in the character of several of the founders—most especially, Robert Boyle, whose colleagues continually praised him for his humility, asceticism, and piety.[87] His story of hearing a calling to God's work at the age of 13 in a thunderstorm in Geneva was a legend even in his own time.[88] If Bacon was the Society's Elijah, Boyle was its Elisha, having inherited a double portion of the prophetic spirit of experimental philosophy. His unimpeachable character deflected from the Society as a whole some of the more inflammatory attacks that critics were willing to aim at individuals.[89] But even the Society's less holy fellows could and did point to its communal practices of witness and validation as guaranteeing its reliability as an oracle of natural knowledge.[90]

These divinatory practices and confirming signs helped the Royal Society secure its authority in Restoration London. But the recognition of this authority—and ultimately, the constitution of its ethos—depended on the polities of the learned who surrounded it and had to be, like Nature, obeyed in order to be commanded.[91] Thus, the dialogues the Society engaged in with early critics in the Church and among its peers were crucial to the establishment of its ethos as a civic oracle.

Prophecy: Old Genres and New Dialogues

When the Society was founded, there was in the air a strong antipathy toward factionalism, since it had been a driving cause of the Civil War. The Clarendon Code (1661–1665) banned religious groups of more than five people from meeting outside the auspices of the Church of England and banished radical preachers from coming within five miles of their former parishes. This was the suspicion facing the founders, who were proposing a Society with lofty religious and political goals and a limited fellowship. To avoid serious political consequences, the Society needed to make as much of its private activity public as was feasible.

At the same time, the Society needed a safe space for experimentation out of the public eye. In an increasingly capitalist economy, it was also concerned about priority for patents for fellows' inventions; likewise, it wanted to secure to England those inventions that might give the nation an edge in military or mercantile competitions. A lively debate erupted in the early 1670s among the fellows about the competing demands of privacy and the Society's propublic charter. Likewise, the Society came under attack from political philosophers such as Thomas Hobbes for the antipatriotic privacy of its operations.

This was the public/private antinomy that faced the Society at its inception. The founders transcended it by engaging critics in prophetic dialogue that kept the Society in conversation with powerful factions in English society—on the basis of shared values such as piety and patriotism—while maintaining political independence. The fellows maintained this dialogue in the *PTRS* and in printed tracts reporting bodies of experiments or rebutting external critiques. These are the documents I now examine for markers of prophetic ethos.[92]

The *PTRS* have been amply scrutinized by rhetoricians.[93] But the first volume, 1665–1666, has been overlooked in much of this work because the genre of the "experimental report" was brand-new, and therefore genre scholars have assumed that there are few stylistic regularities to study during this time. This may be true for some features of the genre. Nonetheless, Gross, Harmon, and Reidy have established some stylistic differences between the early volumes of the *Transactions* and the publications of the French *Académie Royale des Sciences*. These they attribute to the differing social authorization of the two groups—the *Académie*'s role as a court advisory body versus the Society's more independent establishment.[94] Similarly, Dwight Atkinson has noted that the early *PTRS* are "author-centric," exhibiting heavy use of first-person pronouns and active verbs, as in "I measured," "I saw," and so on.[95] Thus, if previous work has judged the representation of ethos and authorization in the early *PTRS* consistent enough to warrant comparisons with the rhetoric of other periods and polities, I believe a more thorough examination of ethos in the first volume is in order.

Accordingly, I have read the 141 reports published in this volume for repeated textual markers of ethos, including personal references and attributions. I also considered the structure of the reports to see if there were any markers of the standard genres of prophetic ethos during this period: aphorism, fable, and judgment formula.

The first volume of the *PTRS* is edited and narrated by the secretary of the Royal Society, Henry Oldenburg. Oldenburg received and presented to the Society letters from fellows and outside correspondents; all experiments reported by correspondence were supposed to be tested or approved by the Society before they were published in the *PTRS*, but that system was not scrupulously observed.[96] Peter Dear has argued that the new authority the Society was developing demanded that the reports in the *PTRS* treat specific, discrete experiences with natural phenomena, and this generalization certainly seems to apply as far as the experimental reports go, but Oldenburg also included many book reviews and untested claims from foreign correspondents in the first volume.

Regardless, Dear points out, and my examination corroborates, that whether articles reported tested or untested natural observations, Oldenburg in every instance went to great lengths to establish the bona fides of his correspondents. This exercise in attribution always dominated the first paragraph of each article. Noblemen, Oldenburg contented himself with introducing via their title, as in "A Relation of Some Extraordinary Tydes in the West-Isles of Scotland, as It Was Communicated by Sr. Robert Moray,"[97] although occasionally Oldenburg would gild the lily on behalf of his favorite founder "that Noble Philosopher, Mr. Robert Boyle." In the case of other correspondents, Oldenburg in *every instance* appended an attribution related to their education, breadth of experience, and/or intelligence the first time they were introduced, and sometimes every time they were mentioned. These epithets included "The Ingenious Mr. Hook," "Monsieur Auzout, a French Gentleman of no ordinary Merit and Learning," "an inquisitive Physician," "an understanding and hardy Sea-man," and "the deservedly famous Monsieur de Fermat." In cases where the source of a report was insecure but Oldenburg wished to print it anyway, he resorted to some serious attributive gymnastics, as in "Notice was lately given by an inquisitive Parisian to a friend of his in London, that by an Acquaintance he had been informed, that Signor Septalio, a Canon in Milan, had lately seen."[98]

Why this obsession with attribution in the early *Transactions*? Right off the bat, we can note that Oldenburg was obeying Bacon's injunction in the *Preparative* to state the source of all natural knowledge as clearly and completely as possible because on the credibility of the source hinged the credibility of the knowledge (4.259–60). Shapin noted Oldenburg's behavior and attributed it to the mores of gentle ethos.[99] Elsewhere, he and Schaffer argued that the style of witnessing that the Society was developing could also derive from legal or "priestly" models.[100] Peter Dear has argued that Oldenburg's obsession with attribution was designed to establish the authority of readings of the "book of nature" against readings of traditional authorities such as Aristotle.[101]

These suggestions are not contradictory: they all deal with the rhetorical function of witnessing, which crucially depends on the trustworthiness of the person who is making her private information public. Nevertheless, I believe Oldenburg's attributions fit into a larger rhetorical pattern that we can see when we consider the body of the *PTRS* reports: Protestant prophetic judgment formula.

Seen from this perspective, Oldenburg's attributions constituted the messenger part of the formula, via which the prophet declared her privileged access to the divine as her authorization to speak. In at least one instance, Oldenburg made the prophetic connection explicit, when he introduced astronomer Adrien Auzout:

> Thus for the Parisian account of the comet, which is here inserted at large, that the intelligent and curious in England may compare their observations therewith, either to verifie these praedictions, or to shew wherein they differ; which is (as was also hinted above) the design of this philosophical prophet in dispersing his conceptions, who declareth himself ready, in case he be mistaken in his reckoning, to learn another hypothesis, to explicate these admirable appearances by.[102]

Two features of this attribution are particularly noteworthy: Oldenburg's use of the epithet "philosophical prophet" to distinguish the kind of divination Auzout is doing from religious divination, and Auzout's stated desire to have his predictions verified by the community. This is one of the first recorded performances of the new hybrid experimental-prophetic ethos.

But even more interestingly, the reports themselves are structured like Protestant divinations. The pattern, remarkably consistent throughout the volume, is as follows:

1. Messenger formula: Oldenburg defends the trustworthiness of the source and the means by which the information was acquired.
2. Exegesis: Some part of the Book of Nature is expounded upon for the benefit of readers: a "monstrous" calf with two heads, the method of mining saltpeter in India, a stone taken from a serpent's head that draws poison from wounds, and so on.
3. Application: In over half the reports, the judgment is specifically applied—either to confirm the Society as a reliable oracle or to suggest a particular benefit to the commonwealth (particularly in industry or medicine).

Thus, all the articles in the first volume adhere to the classical judgment formula, and the majority of them additionally perform the hallmarks of Protestant divination—exegesis from a fundamental textual source (either the Book of Nature or books written by other philosophers) and application to common life. Instead of purely legal or gentle modes of witness, the *PTRS* here borrowed from prophetic genres to convince readers of the authority and utility of the information presented.

Another prophetic genre that appears in the first volume of the *PTRS* is the "inquiry" to far-flung diplomats. These read remarkably like petitions to an oracle: "What river is it in Java Major that turns wood into stone?" "Whether it be true, that diamonds, and other pretious Stones, do grow again after three or four years, in the same places where they have been digged out?" "Whether in Sumatra there be a fountain, running a very Santive Oyl? And whether the ignivomous Mountain in the same Island, do burn continually, and cast out stones so eaten out by the fire, that they swim?"[103] When the replies to these petitions came back months later, they were first read in a Society meeting and then published in the *Transactions*. We see here the founders "crowd-sourcing" witnesses through an oracular petitioning technique and then insisting on communal validation of the reports, much as the Athenians debated the Pythia's pronouncements in the Assembly.

Most of the dialogues the Society engaged in through the *PTRS* were not so formally structured. Continental philosophers such as Auzout, Johannes Hevelius, and Giuseppe Campani corresponded with Society fellows to learn what the English philosophers had discovered or to ask them to corroborate observations. Auzout baldly offered to trade lens-grinding secrets with Robert Hooke.[104] Boyle frequently exchanged techniques and recipes he had discovered, sometimes encrypted, for equally valuable knowledge from his counterparts on the Continent.[105]

But not all of the Society's correspondents were sympathetic to its projects. The two best-documented controversies that embroiled the early Society were fought over its piety and its patriotism with Henry Stubbe and Thomas Hobbes, respectively. Stubbe, a physician and political pen for hire, rebutted Glanvill's *Plus Ultra* (1668) with a pamphlet titled *PLUS ULTRA of Mr. Ioseph Glanvill reduced to a NON-PLUS*. Thus began a protracted pamphlet war that ending up involving Thomas Sprat's *History of the Royal Society* as well. Most of the back-and-forth was meanspirited nitpicking, but the substantive disagreements were as follows: whether experimental philosophy was pious and patriotic or enthusiastic and anarchical; if experimentation could or would ever lead to certain knowledge of natural truth and certain benefit for humanity; and, whether or not the observations of ancient natural philosophers should be used to guide and interpret experiments. Stubbe was unable to offer the kind of evidence the Royal Society could muster in favor of the second and third theses. But his arguments with Glanvill on the first thesis were more damning.

First, Stubbe attacked the Bacon-Boyle thesis that experimental philosophy conduced towards piety, found in both Glanvill and Sprat's publications. Stubbe argued that the Fall had certainly not been occasioned by a failure of natural philosophy and mocked his opponents with a passage from 1 Corinthians (13:2) that denigrated the prophetic ethos they inappropriately wielded: "Though I have the gift of Prophesy, and understand all Mysteries, and all knowledg; and though I have all faith, so that I could remove Mountains, and have no charity, I am nothing."[106] He continued his attack on the founders' prophetic ethos by charging them with heretical enthusiasm, as would Meric Casaubon and Thomas Hobbes.[107]

The Boyle/Hobbes controversy has been admirably detailed by Shapin and Schaffer.[108] I will return to just a few points of their argument. The controversy was over the methods by which the experimenters pursued truth. Hobbes complained of the secrecy of the Society's experiments[109] along the same lines that Stubbe argued: claiming to make knowledge in a privileged polity segregated from the body public both endangered the unity of that body and invalidated any knowledge generated by those means.[110]

The strategies the Society chose in responding to these early criticisms all leveraged prophetic ethos, but in different ways. Glanvill chose to attack Stubbe as a false prophet. In the exhaustively titled pamphlet *A praefatory answer to Mr. Henry Stubbe, the doctor of Warwick wherein the malignity, hypocrisie, falshood of his temper, pretences, reports, and the impertinency of his arguings & quotations in his animadversions on Plus ultra are discovered*—Glanvill put the "prophetik rumper's" attack on the Royal Society in perspective as just one of Stubbe's many well-paid smear campaigns.[111] As evidence along these lines, he dug up old publications of Stubbe written in favor of the Roundheads and against the Church of England.[112]

Unlike Glanvill's ethical assault on Stubbe, Boyle did not accuse Hobbes of being a false prophet of the oracles of church and state. Rather, in his protracted debate with Hobbes—and in related debates with Henry More and Franciscus Linus, Boyle adopted a cluster of dependable ethical strategies identified by Shapin and Schaffer: these included protestations of Boyle's humility and compulsion to publish against his natural modesty; a repeated commitment to "nescience" or lack of belief in any proposition until it be performed and witnessed as a matter of fact; and a "naked" way of writing these observations.[113] According to the authors, these strategies were derived from the strictures of gentle ethos and were aimed at protecting a gentle social microcosm in which consensus could be peacefully built.

By these criteria, Glanvill's "animadversions" on Stubbe are inexplicable—unless Glanvill was viewed as a loose cannon by the Society, and we have no evidence of that sort of censure. However, if we apply the topoi of prophetic ethos, Boyle's and Glanvill's behavior become more coherent: Glanvill was leveling charges of false prophecy against critics to protect the founders' state authorization as a reliable oracle; meanwhile, Boyle was performing humility and compulsion (the "naked" presentation of nature) while relying on the awesome confirming power of his air pump and experimental methods to settle controversies. But most importantly, Boyle and Glanvill were keeping their critics engaged in prophetic dialogue. In this dialogue, as in the ancient oracular consultations, what ended up being important were not the facts at stake—whether the air pump really worked, whether Glanvill was or was not a Papist. Rather, what was key was the articulation of shared values by the disputants. These values—piety and patriotism, tolerance and integrity—kept the disputants bound in polity. As long as the Society could keep its critical peers in conversation, *the conversation itself* legitimated the Society's stake in determining political truth.

Conclusion: The Royal Society as Civic Oracle

In conclusion, the Society exploited the rhetorical resources of prophetic ethos to found a stable and authoritative position that was independent of the body public but remained tied to it via the motivations of Ascertainment, Confirmation, and Prophecy. Through an ongoing dialogue that established its petitioners' esteem for the experimental ethos, the Society helped secure a bulkhead in Europe for science. Within a few decades of the founding of the Royal Society, similar state scientific organizations had sprung up in France (1666) and Germany (1687). And there has been no looking back since. As we move ahead to investigate the prophetic rhetoric of policy scientists in the twentieth century, we see more crucial mutations to the meme of prophetic ethos, but we also find that the ancient markers that survived the upheavals of the seventeenth century persist—because the fundamental questions about how to live with each other and in the world remain unanswered. What Shapin and Schaffer argued at the end of their study still holds: "Our present problems of defining our knowledge, our society, and the relationships between them centre on the same dichotomies between the public and the private, between authority and expertise, that structured the disputes we have examined."[114] And so the prophet remains the first resort in these crises.

5 INTERLUDE

COMPETING ETHICAL MODELS AND A CATCH-22

The hybrid scientific-prophetic ethos as evolved by the early Royal Society plainly expressed the meme of classical prophecy: in the social context, the "cult" of natural philosophy was still established outside politics as usual, to be petitioned for ascertainment in state crises, and the philosophers' special authority was still confirmed via recognition of their miracles and personal performances of calling, asceticism, humility, and compulsion. In textual exchanges, we observed dazzling technical divinations that spurred dialogue between the prophet and the polity, a dialogue that goaded the latter to "know itself" via its political actions.

Nevertheless, significant mutations accrued to the prophetic meme in its hybridization with experimental culture. Namely, the processes of divination became a *communal* enterprise; as with the old saw about bird-watching, no experimental claim counted as the truth unless it was attested by witnesses. Furthermore, the truths divined in this way were themselves no longer divine but *natural*.

In bringing this hybrid meme forward to the twentieth century—the likely focus of most readers' interests and the locus of the second major set of mutations to scientific-prophetic ethos—I face a chasm of 200 years that cannot be filled in by one scholar in one book. So many fascinating crossings of prophetic and scientific ethos lie in the gap: Pasteur's public arguments about invisible microbes, Darwin's theory of evolution, Lyell's calculations of geologic time, Marie Curie's work with radiation. Elsewhere, I have treated a particularly fascinating case from this period—the scientific hoaxes that marked a shift in nineteenth-century America away from humanist and toward scientific ways of deriving social truths;[1] related work has established the era as one of significant hybridization between religious and scientific epistemologies.[2]

Nonetheless, as a compromise between the extremes of leaping blindly over two centuries and leveling the gap with the spadework of many books, I offer a bridge: a rhetorical tension that developed between the progressive model of scientific ethos as practiced by the early Royal Society and a philosophical model of scientific ethos derived from Hume's Guillotine.

Hume's Guillotine and the Prophetic Ethos of Modern Science Advisers

As we have seen, the model the Royal Society founders imagined for the integration of natural philosophy with statecraft was ethically seamless. Natural philosophers would make ideal political strategists; in fact, given world enough and time, they would make excellent philosopher-kings. Who better, after all, to set policy than these Platonic navigators who had sailed past the borders of the cosmos and returned laden with knowledge to be put in operation pro bono publico.

This model of how to integrate scientific and political ethos, which I call the progressive model, persisted into the Enlightenment. By many estimates, it was the very spirit of the Enlightenment, with Isaac Newton figured as its "prophet and patron."[3] A less saintly term for this spirit was Laplace's Demon. This "intelligence" postulated by mathematician Simon LaPlace could, if it knew the present position and momentum of every atom in the universe, attain absolute knowledge thanks to the predictive power of Newtonian mechanics: "nothing would be uncertain and the future, as the past, would be present to its eyes."[4]

"Dare to know!" the Enlightenment philosopher Immanuel Kant exhorted his colleagues in this new epistemological world. Kant here echoed the Delphic motto "know thyself" but with a new twist.[5] Now natural knowledge was marshaled not just for self-definition but toward the total reform of human society. A rational, scientific politics seemed the only panacea for the religious strife of the previous two centuries. Science flourished in liberty and peace; so did republics. Thus began the coevolution of democracy and science in the West. Unfortunately, this process worked about as reliably as Boyle's air pump: we recall that the "scientific" political philosophy that yielded the American Constitution and the philosopher-president Thomas Jefferson also produced the fascist nightmare of Robespierre's Terror.

The year Jefferson's Declaration of Independence was signed was the year David Hume died. His review of Enlightenment proposals for the progressive scientific reform of society was sobering. In a famous section of the *Treatise of Human Nature,* Hume warned his readers that the progressive model had a fatal flaw: policy could not be logically derived from science. Scientific statements were statements about what the current state of the world "is"; policy amounted to statements about what we "ought" to do about the state of the world. The logical gap between "is" and "ought" could not be bridged without the application of something extra: a priori values.

Hume recognized that the gap had been obscured by certain types of policy problems. Say, for instance, scientists observed a meteorite hurtling toward London; the scientists alerted the mayor, and he evacuated the city. This was a case in which the progressive model seemed to operate seamlessly. However, Hume argued, the policy decision actually boiled down to an enthymeme—an argument missing its major premise. "A comet is coming; therefore, we should leave" implied the unstated major premise "meteorites are to be avoided": this was the (quite sensible) value Londoners held that justified the political action. Although this particular value was so universally held as to be practically invisible in the argument, most political values were much less universal and much more contentious: consider, for example, values such as "the state should help out poor people" or "a three-month-old fetus is a human being with full civil rights." These types of values formed obstacles blocking the progressive derivation of policy from scientific knowledge. In fact, *all* values blocked the progressive model, according to the logic of Hume's Guillotine, which has also been called the facts/values, or is/ought, problem.[6]

Science policy expert Roger Pielke Jr. calls political dilemmas such as the meteorite example "tornado politics" and argues along with Hume that these situations have complicated American science policy. They have done so by making it look like all political problems conduce to technical solutions: in other words, we have tended to believe that if we just *had more information,* we could make better policy. But no matter how much data we could lay hands to—even if we were LaPlace's Demon itself—values would still stand in the way of policy making.

The dangers of "tornado politics" were thrown into stark relief during the world wars by the atomic bomb, the Nazi medical experiments, and other technologies of mass killing. To protect society from these perceived abuses of power by science advisers, the is/ought stricture began to be imposed on them.[7] Under the is/ought model, scientists were supposed to figure out what "was" the physical situation in a crisis such as global warming or nuclear war; then, they were supposed to hand off that information to politicians, who were to determine what their polity "ought" to do to solve the crisis.

If the is/ought model seems neat from a philosophical perspective, rhetorically and ethically speaking it is a mess for several reasons—principally because *scientists are citizens*, and as such it is unrealistic to expect them to abjure value judgments about global warming, pesticides, cancer, abortion, or any other issue that affects their families or otherwise impacts their values.

A second problem with the is/ought model of science and policy integration is that scientific work, even "pure" scientific work, tends to proliferate new political dilemmas: think of cloning, genetic modification, or stem cell research as examples of political dilemmas that have arisen on the basis of scientific innovation. Solving these dilemmas generally just leads to the invention of more technologies to "fix" the original problems, and so on. Bruno Latour called this feedback cycle of technologies and dilemmas "hybridization."[8] Hybridization has two codependent phases: the desire to "purify" or insulate society from the vicissitudes of Nature, and the desire to "translate" natural dilemmas into cultural inventions in order to bring nature under human control. So, for instance, the desire to insulate human society from the seasonal binge and purge of harvest and famine led to the invention of agriculture. But agriculture, in fact, just entangled nature and society in a new hybrid configuration: humans stopped wandering in the savannahs and settled in fortified villages; meanwhile, the natural setting changed as certain insect and mammal populations exploded from feeding on the abundant monoculture they found at human farms. These changes in nature put new pressures on human culture—famine again and insect-borne disease—which in turn spurred new attempts to "purify" society from nature via the invention of pesticides. But this strategy amounted to fighting hybrids with hybrids; pesticides soon created their own pressures on culture. This is what the modern project sums to, in Latour's estimation—the proliferation of hybrid technologies and reactionary cultural responses to them.

From Latour's perspective, the is/ought model might count as kind of philosophical purification—the segregation of natural and political knowledge. If so, this model would be predicted to proliferate more hybrids of natural and political knowledge. To wit, the harder we press scientific ethos to stay on the "is" side of the is/ought model, the more pure science we will get. The more pure science, the more new technologies (remember, technologies are not just "hard" like machines but also "soft"—new ways of doing things). The more new technologies, the more policy we need to accommodate them and (since scientists are the only ones who understand the technologies) the more scientists and scientific arguments we need to make policy. To sum up—and this is crucial for understanding the ethical dilemma our science advisers currently face—the implementation of the is/ought model automatically triggers the progressive model, which triggers the is/ought model as a reaction, and around and around we go.

This is the philosophical problem plaguing our twentieth-century science advisers, but how does it cash out rhetorically, in words and deeds? We can see the entanglement of "is" and "ought" arguments in public debates about science if we apply the lens of stasis.

Stasis Theory and Scientific-Prophetic Ethos

Stasis theory is a simple and elegant way to expose the internal philosophical structure of public argumentation. The *stases* were derived by Roman orators from judicial arguments; they indicated "sticking points" on which consensus had to be reached before civic action could be taken. They were organized stepwise because stasis also meant "level." The stasis of fact addressed the question, What happened here? Once the facts were in hand, the jury could ask, How do we classify such happenings? (Definition); once the jury had categorized the event, the question arose, How do we evaluate this event? (Value); and finally, How should we proceed from here? (Action).[9]

This sequence still dominates our current judicial process—or our deliberative legislative processes, for that matter, as well as our committee hearings, department meetings, disciplinary discussions with our children, and so on. The stases are an extremely durable meme in Western polities because they are a rhetorical technology for organizing action in a crisis among free agents. The stases are irrelevant in a totalitarian regime or a theocracy, in which there are very few free agents—thus, no debate and no "sticking points." But any polity where cooperation must be solicited and organized on a large scale needs some system like the stases.

In scientific polities, Jeanne Fahnestock and Marie Secor found that the four classical stases of argumentation became five stases in practice: the stasis of definition subdivided into questions of classification and questions of cause and effect.[10] So, if we combine this result with Hume's original conception of the is/ought problem, we can see that the stases of fact, definition, and cause/effect fall on the scientific or "is" side of the divide while the stases of value and action land on the "ought" or policy side.[11] According to this model of scientific ethos in policy making, scientists should ascertain the facts of a phenomenon (e.g., the seismic readings of an earthquake), classify them (e.g., the earthquake ranks 7 on the Richter scale), and then use that classification to determine as far as possible the causes and likely effects of the phenomenon (e.g., the earthquake is likely to produce significant aftershocks). At this point the scientists' work is done; they pass their information to politicians to evaluate in terms of voters' expressed values (e.g., we value our lives more than our business and our daily routines), and these value judgments on the data lay a certain foundation for political action (e.g., evacuation of the city in anticipation of aftershocks).

I have already reviewed the psychological and sociological reasons that such a hand-off proves impracticable most of the time. Stasis theory provides a rhetorical explanation of the failure of the is/ought model: that is, *the stases exert an "upward" pull toward action on all public scientific statements*, particularly on the connotations of technical definitions and on the effects scientists judge likely. This upward pull is generated by the predictability of the stasis meme in Western polities. We all have daily and repeated experience with the stases in public argument, and so we all know what's coming when debate over a dilemma begins. In chapter 10, we see the chilling effects of this upward pull in the trial of Italian seismologists for their alleged failure to sufficiently warn the public about an impending earthquake. But here is another brief example: Congressional hearings on regulating nicotine in cigarettes in the mid-1990s turned on the stasis of whether or not nicotine was addictive. This was a definitional question, not a question of value or action; nevertheless, the word "addictive" was heard to imply a value judgment. We could anticipate that Congress was going to make the value judgment that addiction was a bad thing and, therefore, that action should be taken to warn or protect the public from it—and in fact, that was exactly the result with the Tobacco Master Settlement in 1998.

A few more examples: in debates over the fluoridation of public water supplies, the fact that fluoride was classified as a "chemical" triggered public resistance against the effort because, in the wake of fallout and pesticide scares, audiences equated "chemical" with "poison" (even though, as we all learned in high school chemistry, most everything we ingest is strictly speaking a chemical, including water itself).[12] Recently, Andrew Wakefield's claims (proven false) that mercury in some doses of the mumps-measles-rubella vaccine appeared to be one cause of autism were interpreted by concerned parents as arguing that all childhood vaccines were dangerous even though Wakefield himself never made that value judgment.[13] Finally, scientists have been accused of political bias when they draw conclusions that some or another industrial product causes cancer.[14]

To sum up, in all policy debates *we know from the beginning that we are headed toward value judgment and policy*. For this reason, lower-stasis arguments are frequently heard as higher-stasis than they really are. In science policy debates, claims that scientists make about the state of nature are heard as value judgments and recommendations—despite the fact that the claims *do not contain value- or action-specific language*. This upward pull is the deciding factor in the dysfunctional ethos of the American science adviser, but to understand exactly why, we have to look more specifically at the way the competing is/ought and progressive models for appropriate scientific ethos developed in the early twentieth century.

Competing Models of Prophetic Ethos in the Early Twentieth Century

Historians and sociologists in the first half of the twentieth century were not unaware of the problems created by the interaction of the is/ought stricture with scientific ethos. Max Weber, who famously derived the "spirit" of capitalism from the Puritan work ethic, also believed that the scientific ethos had its roots in ascetic Protestantism. But since the Enlightenment, he argued in "Science as Vocation," science had lost its traction on moral issues: "Tolstoi has given the simplest answer, with the words: 'Science is meaningless because it gives no answer to our question, the only question important for us: "What shall we do and how shall we live?"' That science does not give an answer to this is indisputable."[15] He went on to argue, as Einstein would argue 40 years later, that the best and only way science could serve society was by providing an apparatus for scrutinizing policy questions and for enacting policy.[16] This argument was essentially a reformulation of the is/ought model.

The scientific ethos predicated on these ethical strictures was most fully described by sociologist Robert K. Merton. It consisted of four ethical norms, often referred to by their acronym CUDOS: Communism, meaning scientific knowledge is the property not of the individual scientist but of the entire scientific community; Universalism, denoting the independence of scientific truths from the cultural context of their discovery; Disinterestedness, denoting the freedom of scientific work from personal bias; and Organized Skepticism, denoting the habitual submission of all scientific claims to rigorous testing.[17] These norms put pressure on scientists to avoid claims that would reveal personal political beliefs or that could be perceived as pushing policy. That the norms were powerful in the pre-war period (and persisted thereafter) has been attested by multiple sociological studies performed since the first formulation of CUDOS.[18] These scholars have adduced evidence that the norms are espoused by scientists in many fields and that deviation from the norms (say, taking industry money for pesticide research or patenting a technology so that it becomes individual rather than common property) is met with strong criticism—even though such deviations are in fact commonplace. Thus, while CUDOS is not an accurate description of scientists' actual behavior, it nevertheless has proven to be a powerful stereotype both within and outside scientific polities.

The persistence of the Mertonian stereotype in the first half of the twentieth century is corroborated by contemporary popular portraits of scientists in news media and in cultural works—too many to catalogue here but epitomized by Sinclair Lewis's novel *Arrowsmith* (1925). Martin Arrowsmith is a scientist so dedicated to his work that he disregards society, politics, and personal safety. His mentor's name, Gottlieb, translates "Love of God." If this personification of monastic devotion to the ethical norms of CUDOS strikes us as over the top, it was nonetheless right in line with contemporary commentary, which declared scientists "zealous," "passionless," "devoted" "moral eunuchs or civic hermits" who, in Julian Huxley's estimation, were "mak[ing] possible the emergence of a more universal and more adequate religion."[19]

But even as this monastic Mertonian ethos, which I will call "disciplinary ethos" from now on, was gaining cultural currency, the progressive model of scientific-prophetic ethos experienced a resurgence. This Renaissance was heralded by an essay entitled "Scientific Research: The Art of Revelation and of Prophecy," published by mechanical engineering professor Robert Henry Thurston in 1902.[20] Thurston was a forerunner of the Technocrat movement founded by Thorstein Veblen and Howard Scott two decades later. Technocrats contended that scientists, by virtue of their privileged access to natural knowledge, were the best people to determine policy in all matters. Intellectual heirs of Jefferson and the Enlightenment, the Technocrats included men such as President Warren Wilson and the Beard brothers, historians who penned a magisterial survey of American government titled *The American Leviathan: The Republic in the Machine Age* (1930).

The Technocrats drew sharp criticism from adherents to the is/ought model; these critics argued that it was dangerous to place so much power in the hands of an elite group.[21] President Dwight D. Eisenhower would echo this criticism nearly 30 years later in his own prophetic warning against letting the "scientific-technological elite" embedded in the Executive, particularly the "military-industrial complex," take decisions out of the hands of voters.[22] But even these critics had to admit there was no way to make modern policy without scientific advice. That bridge had long since been crossed.

The Technocrat and the Prophet

The rise of the cult of expertise in interwar America calls for a few words on the way the term "expert" has developed in the scholarship on science advisers and on how expertise is related to prophetic ethos.

Scholars agree that the "expert" is a relatively recent ethos. The main factors in its rise appear to be the growth of massively urban civilizations, the resulting breakdown of personal trust as a guarantor of reliable information, and the attendant rise of technologies of standardized production.[23] Max Weber treated expertise as a function of rational authority. As I noted in chapter 1, Weber argued for three major epochs in the historical development of authority or leadership ethos: charismatic, traditional, and rational. Charismatic authority, like the Greek root from which the term derived, was wielded on the basis of a "gift of grace" from a supernatural power. As the functions of charismatic authority became routinized, it morphed into traditional authority of the type we often call king- or chiefdom. Instead of being guaranteed by supernatural donation, traditional authority was guaranteed by hereditary nobility. With the development of more complex and technical polities in the Enlightenment—republics and democracies—traditional authority was routinized further and became a function of due process: rational authority. Rational authority was and is guaranteed not by God or by lineage but by procedure, and so in principle, any individual can wield it.[24]

Historian of science Peter Galison calls this emphasis on procedure in the construction of expert ethos "mechanical objectivity." Mechanical objectivity became associated with scientific practice in the nineteenth century and saw its heyday in the Technocratic era, when it helped create "neutral, self-abnegating bureaucrats" who were "determined to become transparent to nature."[25] As Theodore Porter has pointed out, science advisers still justify their recommendations with statistics and other quantifications because by doing so, they perform mechanical objectivity; in other words, quantification is proof against individual bias.[26]

But despite its appeals to openness and transparency, mechanical objectivity is not necessarily democratic. Procedures such as risk analyses and economic projections may have helped guarantee trust among the experts who wielded them, but most Americans cannot understand these techniques without years of special training and the money to purchase that training. So, rather than serving as classical evidence in a public debate, the techniques of mechanical objectivity have often proved as legible as a pile of oracle bones on the floor or the ravings of the Pythia; they amount to another confirming sign of the charismatic authority of science advisers. No wonder the cult of mechanical objectivity quickly led to warnings against "doctor worship" and the dominance of a "scientific-technological elite."

After the Progressive era, Galison records one more epoch in the development of technical authority—a turn toward the "interpretive." Since the 1930s and 1940s, the "'trained expert' has come to designate not so much an initiate into a secret set of skills, but a potential 'everyman' who can come through training to wield the 'experienced eye.'"[27] In support of this thesis, Galison points to the midcentury proliferation of medical atlases and correspondence courses designed to make expert practices legible to and imitable by the lay public. Porter adds to this lot university training courses: "There the ineffable skill of the craft or build is, so far as possible, made formal and explicit, and thus the secrets of the trade are deemphasized."[28]

However, Porter goes on to say that in practice, all these arguments for "trained expertise," for democratic access to technical authority, are just that—arguments, not social facts. Sociological studies of trained expertise reveal that "expert knowledge is almost by definition possessed by only a few, and no such art is ever reduced to a handful of rules that can be looked up and mastered by anyone with a textbook. Thus the intuition or judgment of specialists continues to command a degree of respect."[29]

We see persisting through these changes rung to the constitution of "expertise" a bevy of prophetic confirming signs: calling, asceticism, and humility chief among them. True, in the twentieth century, "vocation" becomes a matter of social class and specialized training rather than revelation—hair shirts are swapped for spreadsheets—but to the extent that these performances are beyond the reach of the majority of Americans, to that extent the meme of prophetic ethos persists in the ethos of the science adviser.

Here, then, is the difference I draw between prophetic ethos and expertise in the rest of the book. To the degree that a particular skill or perspective is presented as attainable by the majority, to that degree I say expertise is operating. To the extent that knowledge or technologies are presented as beyond the ken of the majority, I find prophetic ethos being performed. A science adviser may perform both expertise and prophetic ethos—to the same audience or to different audiences; we will encounter a great deal of such interaction in chapter 8. Nevertheless, I think it is important to distinguish performances of democratic and rational expert ethos from performances of charismatic and undemocratic prophetic ethos because the problems—and therefore the solutions—that arise from these two performances differ. Claiming prophetic ethos means claiming a privileged and knowledgeable stance outside politics as usual. Policy claims (claims at the stases of value and action) articulated from this rhetorical stance are therefore prophetic claims. For science advisers, then, the philosophical dividing line between "is" and "ought" maps to the rhetorical dividing line between disciplinary ethos and prophetic ethos.

Prophetic Ethos and Scientific Ethos Leading into World War II

Having reviewed the rise of the Technocrats, we might reasonably expect that they embraced prophetic ethos while the is/ought camp forswore it. However, as it turns out, spokespersons for both models made their cases in strongly prophetic terms.

Defenders of the progressive model wanted scientists placed as high in the Executive as possible, ideally overseeing all American industry via a "National Directorate."[30] Technocrats made strong predictions for the collapse of the modern price system and a dystopia of unemployment.[31] There was a lot of millennialist talk, as well, flying under the old Victorian banner of "Progress": science and technology would end disease, hunger, war, poverty, back-breaking labor; they would make America a land of milk and honey.[32] Of course there were risks in placing all of America's eggs in one ethical basket, but the safety net was to be found in scientists' own selfless devotion to the public good. After all, they sought neither fame nor wealth, but only the truth.[33]

Much of this argumentation could have been ripped straight from Francis Bacon or Thomas Sprat's notebooks. No surprise there, as Francis Bacon's ideal natural philosopher looked a lot like Veblen's Technocrat. What comes as a bit of a shock is the intensely prophetic rhetoric of the is/ought advocates.

Then again, perhaps it is not so surprising. If we take another look at Merton's norms—Communism, Universalism, Disinterestedness, and Organized Skepticism—and remember their apotheosis in Martin Arrowsmith, we are reminded of nothing so much as Robert Boyle.[34] We get the same result if we revisit Weber's arguments about the scientific vocation. As I indicated earlier, Weber argued that the prophet had no place in scientific research and teaching.[35] But his definition of the proper scientific ethos nevertheless ticks most of the boxes of the prophetic-scientific meme: independence from politics yet loyalty to the state, a strong sense of "calling," asceticism, humility, and above all, the charisma and duty to engage citizens in dialogue that convicts them of their moral obligations.[36]

We can make sense of this seeming paradox in Weber's thought if we entertain the premise that he attempts to purify prophetic ethos of demagoguery and co-opt the rest of its charismatic authority for science. Indeed, the ways Weber talks about "prophets" are the ways we usually talk about demagogues: they abuse charisma and technical skill to intimidate or dazzle audiences; they leverage expertise in one field to lend a false aura of legitimacy to claims in all fields.[37] All the good qualities of prophets Weber attributes to his ideal scientists. We can appreciate this rhetorical hair-splitting if we put Weber in his proper context. During World War I, he watched many of his scientist friends in Germany take vehement nationalist positions to devastating political effect. It thus makes sense that he would be extraordinarily sensitive to abuses of charismatic authority. Einstein made remarkably similar prescriptions after similar experiences in World War I.[38] But, ironically, Weber's and Einstein's attempts to enforce the is/ought model among their colleagues wound up subverting it. By commanding their scientific peers to engage the public in ethical dialogue, Weber and Einstein were exhorting them toward a stance from which their pronouncements would be interpreted as prophetic.

So, as science advisers entered the public arena en masse after World War II, their behavior was subject to interpretation under competing ethical models—one that put pressure on them to guide America to a brighter future, and one that instructed them to hand over their discoveries but keep their political opinions to themselves. The is/ought model placed an additional constraint on science advisers due to the upward pull of the stases: namely, claims that science advisers made about causes and effects, though technically on the "safe" disciplinary side of the is/ought divide, were nonetheless heard as prophetic value judgments and recommendations—especially when they concerned touchy definitions or frightening effects. These perceived transgressions of the is/ought boundary were predictably met with attacks on advisers' disciplinary ethos—accusations of fraud, bias, and other violations of Mertonian norms. In other words, their polity was telling them, "You can't be both a politician and a scientist, and since you're acting like a politician right now, you must not be a scientist." We see this catch-22 operate in all the twentieth century cases to follow, and it remains in force whenever the is/ought model is the dominant model in science policy making.

But these memes inherited from the Enlightenment and the Progressive era did not singlehandedly set the ethical scene in which postwar science advisers operated. Naturally, kairos added its own pressures.

Science in Postwar America

All science advisers in the West have inherited prophetic ethos, but the ways that ethos is integrated in the policy-making process today depend on each polity's civic epistemology—its preferred way of sourcing and validating public truth. From here on, I follow the American evolution of the meme of scientific-prophetic ethos, though I draw a few comparisons with British and German civic epistemologies in chapters 8 and 9.

It's standard practice to argue that the atomic bomb profoundly altered the public awareness of and relationship to scientists in the United States.[39] But for every argument that scientists suddenly became public saints (or sinners) on August 6, 1945, we can find a commentator arguing along the same lines in the interwar era.[40] Josiah Stamp even mounted a public campaign to halt scientific research until society could "adjust" to the current panoply of technologies.[41] In fact, criticism of the social impacts of science and technology can be traced in the American tradition at least back to Hawthorne's and Melville's stories "Ethan Brand," "The Birthmark," and "The Tartarus of Maids."[42] Truly, the unstable ethos of the science adviser during the first half of the twentieth century owes as much to long-standing public anxiety about the risks of technology as it does to the tension in ethical models I outlined above.[43]

So, if the atom bomb did not necessarily revolutionize the relationship between American scientists and their publics, it certainly intensified the oscillations between "saint" and "sinner" in popular portrayals of scientists. On the one hand, scientists had won the war, terribly and mystically. Unlike the "gears and girders" technology of the steam age, where nearly any viewer with enough time and attention could roughly make out "how" a particular weapon worked,[44] atomic explosions were created invisibly. The scientists who understood atoms and could marshal them into such destruction seemed little short of Old Testament prophets calling down fire from a clear blue sky. Nothing seemed impossible for the "sacred fraternity" of "modern Prometheans," "wizards," "magicians," and "high priests in the temple of Science"; Father Daniel Dubarle wrote in 1956 that "humanity of today is in far-reaching agreement with the secular prophecies of the scientific mind."[45] Popular and serious accounts alike anticipated scientific utopias.[46]

But at the same time, and increasingly, many postwar political and literary portrayals of scientists presented them as dangerous Dr. Fausts or Dr. Frankensteins.[47] The stories filtering back after the war about the atrocities committed by Nazi scientists reinforced these impressions; so did the mounting evidence for the health risks of fallout despite scientists' assurances about the safety of nuclear weapons tests[48]—which starting in 1961, proceeded at the astonishing *average* rate of 27 shots a year for three decades.[49] Philosopher of science Karl Popper noted in a speech toward the end of this decade that the atom had shown Americans that even the most esoteric scientific pursuits could have devastating applications. As a result, scientists had a moral responsibility: *sagesse oblige*. Those with privileged access to knowledge had to take responsibility for the technologies created with it.[50]

If the wider constitution of scientific ethos seemed to oscillate unpredictably after World War II, the governmental reception of scientists postwar was, initially at least, solidly positive. Atomic scientists had produced a powerful confirming sign both of their natural authority and of America's military dominance. Many of the workers on the Manhattan Project went from laboring unmolested in their laboratories to sweating under the klieg lights of congressional inquiries into everything from the Soviet progress in atomic energy to the possibilities of nuclear-powered electricity and medicine—all policy debates to which scientific advice was indispensable. But after a while, politicians and analysts began flinging accusations of "doctor worship" at each other and worrying that no one in Washington actually understood enough of what the "doctors" were saying to make effective atomic policy decisions.[51] One of our first science policy historians, Don K. Price, put the dilemma this way:

> [T]he plain fact is that science has become the major Establishment in the American political system: the only set of institutions for which tax funds are appropriated almost on faith, and under concordats which protect the autonomy, if not the cloistered calm, of the laboratory. The intellectual problems involved in this new status are likely to trouble scientists almost as much as the fears of the apocalyptic uses to which their discoveries may be put by the politicians.[52]

This was the political dilemma that led Eisenhower to warn about overreliance on the "scientific-technological elite." The primary source of Eisenhower's anxiety was the contractor relationship that had produced most of America's weapons research and development (R&D). For reasons of efficiency and secrecy during the war and the following Cold War, Congress had begun making massive R&D appropriations to private corporations, and these corporations hired droves of scientists, either directly or as subcontractors on federal grants to universities (the National Science Foundation did not exist until 1950 and was not in a position to distribute serious amounts of federal funding until several years later). The names of these corporations should sound familiar: Lockheed, Boeing, Raytheon, Bell Laboratories. By the mid 1950s, over half of the nation's R&D budget was doled out to the private sector. A big chunk of these appropriations went to national defense, as indicated above, but huge sums were also channeled to agrichemical and pharmaceutical R&D by a similar mechanism.[53] The term for this mechanism was Big Science, coined by physicist Alvin Weinberg in a startlingly cultic reaction to President Eisenhower's farewell address:

> When history looks at the twentieth century, she will see science and technology as its theme; she will find in the monuments of Big Science—the huge rockets, the high-energy accelerators, the high-flux research reactors—symbols of our time just as surely as she finds in Notre Dame a symbol of the Middle Ages. She might even see analogies between our

> motivations for building these tools of giant science and the motivations of the church builders and the pyramid builders. We build our monuments in the name of scientific truth, they built theirs in the name of religious truth; we use our Big Science to add to our country's prestige, they used their churches for their cities' prestige; we build to placate what ex-President Eisenhower suggested could become a dominant scientific caste, they built to please the priests of Isis and Osiris.[54]

Scientists coming out of graduate programs increasingly found the best jobs in industry; along with the plump salary, kitted-out lab, and comfortable swivel chair, these jobs frequently came with tedious federal security protocols and office wiretaps. At least two surveys of young scientists in the 1950s found a significant proportion weighing the salaries of Big Science against the scrutiny of Big Brother.[55]

When these scientists' federal employers summoned them to Washington, there was little doubt why: it was to make predictions.[56] In the case of the atomic scientists, they were summoned to make predictions about when the Soviets would create their first A-bomb, predictions of the effects of fallout, predictions about the capacities of the world uranium supply, predictions of the public health effects of a nuclear war. These science advisers were tasked with ascertainment in what had just become a radically uncertain global political landscape. To model this landscape, mathematicians and statisticians developed systems analysis. But in terms of gaining certainty, this strategy was like swallowing a spider to catch a fly: Congress asked for certain predictions about Soviet nuclear capacities, for example, and what they received were more uncertainties in the form of model parameters and statistical ranges. Scientists were constantly being pressured to remove the uncertainty from their testimony and to make strong claims about future effects on which grounds policy decisions could be based.[57] Thus, the pressure toward prophetic pronouncements was extreme. Whether scientists were censured for these pronouncements depended on which administration was in power, and therefore which model of scientific-prophetic ethos—the progressive or the is/ought—was in currency.

Conclusion: The Fractured Ethos of the American Science Advisor

This chapter has reviewed the major factors influencing the construction of late-modern scientific ethos in light of my findings so far concerning the persistent meme of prophetic ethos. We have seen that at least two dominant models for the public behavior of scientists competed with each other in postwar America—the progressive and is/ought models. The progressive model, rooted in the Royal Society's practices, posited that since scientists best understood the world, they were best positioned to set policy. This model was most dominant in the interwar period but experiences a resurgence whenever a progressive administration comes to power. The is/ought model tends to be activated either by conservative administrations suspicious of centralized regulation or in response to perceived abuses of power by scientists. But neither model ever goes away entirely, and the latent competition between them creates an ethical catch-22 for science advisers in the following way: in policy crises, the old progressive model triggers calls to scientists for advice. Some science advisers respond with bold policy recommendations. Others, mindful of the is/ought stricture, try very hard to restrict their advice to definitional or causal claims—to no avail, as the upward pull of the stases ensures these claims will be heard as prophetic value judgments anyway. Whichever tack they choose to take, science advisers are frequently punished—typically by conservative or reactionary factions in government—for having transgressed the is/ought divide; thus, the catch-22. This ethical punishment generally takes the form of attacks against the scientists' disciplinary ethos—the very expertise that warranted their calling as public advisers in the first place.

Science advisers risk triggering these ironic attacks *every time* they address evaluations or recommendations to nonscientific polities; when they address scientific polities, the situation is more nuanced, as I explain in chapter 8. But in any event, it's not hard to see that the bully pulpit of the science adviser has a crack running down the middle. This instability poses a risk to anyone who steps up into it—as Robert Oppenheimer learned the hard way.

6 J. ROBERT OPPENHEIMER
CULTIC PROPHET

From the air, Los Alamos and Delphi bear an eerie resemblance (figure 4), and I would argue the resemblance is more than coincidental. Like Delphi, the Manhattan Project was located far outside the cosmos of national politics to ensure its security. The American government authorized its workers as long as they performed signs confirming their privileged access to natural knowledge and observed a rigid asceticism—long working hours, fences, background checks, wiretaps, interrogations, censorship, and curfews.

Like Delphi, Los Alamos was founded to ascertain matters of national security—in this case, whether atomic energy could provide a way out of the quagmire of the Pacific theater of World War II. The scientists engaged in this ascertainment provided an apocalyptic response to the government's petition. But in ending the war, the bomb ushered in an entirely new world—one in which entire peoples could be incinerated with the push of a button.

The explosions at Hiroshima and Nagasaki made horrifically and awesomely public the previously secret atomic divinations of the scientists on "The Hill." They brought these men into prophetic dialogue with generals, executive officials, elected representatives, readers of national newspapers, and television viewers. Everyone wanted to know what the chances of nuclear holocaust were and what its effects would be. Oppenheimer, as scientific director of the Manhattan Project, became its spokesman,[1] like a *prophētēs* chosen to inscribe the Pythia's divinations. In this cultic role, Oppenheimer discovered the unpredictability of administrative tolerance for scientist-prophets. Examining the constitution of his ethos during his security trials in 1954, as well as his rehabilitation by the Kennedy administration in 1963, will reveal how deep the fractures in the ethos of the science adviser can run.

(A)

(B)

FIGURE 4 (A) Aerial photo of Delphi, Greece, taken in the early 2000s. The oracle itself was located to the right of the current town settlement, where ruins are visible at the foot of the Castalian cleft. (B) Aerial photo of Los Alamos National Laboratories, Los Alamos, New Mexico, taken in the 1970s.

Source: Photos courtesy of Yiannis Koulelis (A) and Los Alamos Historical Museum Photo Archives (B).

Los Alamos

J. Robert Oppenheimer was recruited from the University of California, Berkeley, to run the Manhattan Project in 1942. He had worked under Niels Bohr and had published a few important studies from a quantum-theoretic perspective on astronomy, spectroscopy, and electrodynamics. As his student Joe Weinberg succinctly framed Oppenheimer's professional standing at the time, "Bohr was God; and Oppie was his prophet."[2] Weinberg was one of a group of Oppenheimer disciples who hung on the physicist's every word and even went so far as to copy his dress and mannerisms.

It was this intense personal charisma, and a knack for explaining complex technical problems in lay language, that brought Oppenheimer to the attention of General Leslie Groves and the federal agents in charge of building "the Gadget" at Los Alamos. In fact, Oppenheimer suggested the project site, having become so enraptured with the isolation of the area several years earlier that he had taken a long-term lease on a ranch across the valley.

He quickly proved himself a capable administrator. Although those duties and the constant pressures of liaising with the military officials in charge of the project kept him from making significant scientific contributions to the fission bomb—the heavy lifting here was done by Enrico Fermi, Richard Feynman, Hans Bethe, Edward Teller, and others—Oppenheimer had a universally acknowledged knack for taking scattershot discussion on a technical issue and synthesizing it into a coherent and soluble problem statement.[3] In that way he was crucial in moving the scientific work forward.

The ascetic lifestyle at Los Alamos reduced Oppenheimer to a scarecrow physique and a constant, racking cough. No one knew for sure whether or not the Trinity test would ignite all the free oxygen in the atmosphere: Fermi started a betting pool on it the night before the test. When the clock finally ticked to zero on July 16, 1945, the Jornada del Muerte melted into a searing blur of white. As Oppenheimer's eyes adjusted behind his dark glasses and protective screen, he heard the delayed thunder of the blast and watched the great violet cloud unfurl its wings over the mountains. Later, he would claim he remembered at that moment a passage from the Hindu spiritual epic *Bhagavad-gita*: "Now I am become death, the destroyer of worlds."[4]

That apocalyptic vision also exploded Oppenheimer's moral and political horizons. Previously confined to the narrow polity of Los Alamos, they now encompassed the globe. He pleaded with his military supervisors to give some advance warning of the planned attack—if not to the Japanese themselves, at least to the United States' Russian allies—but to no avail. The detonations over Hiroshima and Nagasaki, on August 6 and 9, respectively, turned him into a "nervous wreck," according to FBI agents monitoring the bugs in Oppenheimer's home.[5] By November of that year, he was lobbying his powerful friends in the Truman administration for civilian and international oversight of atomic energy development. His last speech as director of Los Alamos, on October 16, 1946, was riddled with jeremiads: "If atomic bombs are to be added as new weapons to the arsenals of a warring world, or to the arsenals of nations preparing for war, then the time will come when mankind will curse the names of Los Alamos and Hiroshima" and "The peoples of this world must unite or they will perish. This war, that has ravaged so much of the earth, has written these words. The atomic bomb has spelled them out for all men to understand."[6]

By the time he gave this speech to his colleagues, Oppenheimer had already begun making the same sorts of oracular statements to Congress. For example, in his testimony before the McMahon committee on Atomic Energy in December 1945, he uttered a litany of prophetic aphorisms that read like Bacon's *Sylva Sylvarum* updated for the atomic age:

> There are, and there will be, no specific countermeasures to atomic weapons....
>
> There can be no long term security, either in the scientific novelty, or in the vast technical complexity, of the project....
>
> Atomic weapons can now be, and could increasingly be, cheap instruments of devastation, cheap compared to the costs of war as we have known them.
>
> Atomic weapons are now, and could increasingly be, effective instruments of devastation, so effective that they are unlikely, in a world at war, to be renounced for humane or for tactical reasons.[7]

It was clear that he was starting to relish the bully pulpit of the prophetic science adviser. Over the next five years, he was appointed to eight executive and legislative advisory committees, many of whose deliberations replicated the secrecy of the inner sanctum on the Hill. His most important position was as chairman, from 1947 to 1952, of the Atomic Energy Commission's (AEC's) scientific General Advisory Committee (GAC). Oppenheimer served on the GAC with luminaries such as Fermi and James Conant, the president of Harvard. A vault stood outside his office to protect classified documents, guarded round the clock by armed officers.

These heady responsibilities were quickly reflected in Oppenheimer's public rhetoric. He began peppering testimony and speeches with

I-could-tell-you-but-then-I-would-have-to-kill-you-type statements that elliptically confirmed his privileged access to natural and state secrets.[8] He was spending so much of his time on planes between Cal Tech/Berkeley and Washington, DC, that he accepted an invitation to head the Princeton Institute of Advanced Studies (IAS) principally to shorten his commute.

He had also become a media celebrity. His face, dominated by his wide-set, otherworldly blue eyes and frequently haloed in quantum equations (figure 5) stared out at readers from the cover of every major news weekly in the country at some point between 1945 and 1950. He gave dozens of interviews and speeches in which he pontificated on a stunning range of political, social, ethical, literary, scientific, and philosophical topics. He developed a reputation for "oracular" aphoristic pronouncements on these subjects,[9] the most famous being "the physicists have known sin; and this is a knowledge which they cannot lose."[10] The wider polity he reached through mass media apparently both recognized and reinforced the prophetic ethos Oppenheimer was constructing: reporters and writers of letters to the editor frequently characterized him in priestly or prophetic terms.[11] Many of his mass media messages differed starkly in their pacificism from the war counsel he gave in executive sessions.[12] And so he developed a split prophetic personality: the cultic war adviser and the kairotic pacifist crying out in the wilderness.

FIGURE 5 J. Robert Oppenheimer (right) posing with mathematician Oswald Veblen at the Princeton Institute for Advanced Studies. Oppenheimer was frequently photographed in close frame with a "chalkboard halo" of physics equations behind his head.

Source: Courtesy of the Associated Press.

In cultivating these two prophetic ethē, Oppenheimer had created an unstable ethical mixture. "He wanted to be on good terms with the Washington generals... and to be a savior of humanity at the same time," concluded his student Freeman Dyson.[13] "He walked along the edge of a precipice," worried Oppenheimer's friend and fellow atomic physicist Isador Rabi. "He didn't pay attention to the outward symbols."[14] This tightrope act was doubly tense given that conservatives were enforcing the is/ought model as a way to control the flow of scientific information in the opening years of the Cold War. And the rope finally snapped in June 1954 when the AEC revoked Oppenheimer's national security clearance *one day* before it was due to expire, excluding him from policy making for the rest of his life. We now delve into that kairos, focusing on how Oppenheimer and his opponents constituted his ethos with respect to the is/ought model and the topos of security.

Security and Certainty

Security was the dominant topos of postwar politics: it underpinned the Red Scare, escalated the Cold War, and imposed rigid controls on the public and private behavior of scientists who took federal money for research. Security is intimately related to Ascertainment, the purpose of prophetic dialogue. If political certainty, as I have constructed it, denotes judgments at the stasis of value, security is the policy face of that certainty—arguments made at the stasis of action to maintain certainty over time and in the face of unwanted change. Thus, the United States has food security if its people eat well in times of drought; it has nuclear security if its atomic stockpiles stay where they are in spite of attempts to seize them. Security is measured negatively: nothing lost, nothing stolen, nothing changed.

Given my definition, it stands to reason that if a polity calls on a prophet to help it ascertain what to do in a given crisis, it will also seek to secure that channel. And indeed, this is one of the main functions of Authorization—to set up the prophet in a safe location, to restrict access, to guard the prophetic office against bribery, coercion, defection. All these security measures obtained in spades at Los Alamos. They were more difficult to carry out when Oppenheimer was off campus; hence, the wiretaps, the cars parked outside his lovers' apartments, the undercover agents sitting behind him on his plane and train commutes. But perhaps the best guarantees of security were the personal habits Oppenheimer had internalized at Los Alamos. He would safety pin classified memos inside his suit coat pocket so they wouldn't fall out or get stolen on the train. He was aware of a good percentage of the FBI's attempts to keep track of him, legally or otherwise. His habits were so fastidious in this respect that investigators in his clearance hearings would be unable to prove that Oppenheimer had leaked so much as a single character of classified information during his decade of federal security clearance. The campaign to "unfrock" Oppenheimer as the government's chief atomic prophet would have to turn on other points.[15]

Oppenheimer's Security Clearance Hearings, 1953–1954

In 1953 Oppenheimer was near the zenith of his public career. He was Einstein's boss at the IAS; he had just wrapped up a distinguished term as GAC chairman, receiving a personal thank-you note from President Eisenhower for that service, and he was still consulting for the AEC. His Reith lectures for the British Broadcasting Corporation (BBC) had just come out in book form; and, he was still a regular feature of the news weeklies. But with his public reflections on the risks of atomic energy, he had made some very dangerous political enemies, chief among them Admiral Lewis Strauss, the chairman of the AEC and Eisenhower's atomic energy adviser. Strauss was a war hawk who resented both Oppenheimer's public pacifism and the physicist's personal manner, which Strauss perceived as insufferably arrogant. The admiral dedicated pages of his private journal to a litany of sleights he felt Oppenheimer had dealt him during AEC deliberations.[16]

One of the most unforgiveable of these challenges, from Strauss's perspective, was a speech Oppenheimer gave to the Council on Foreign Relations in February 1953, reprinted in *Foreign Affairs,* and in the *Bulletin of Atomic Scientists* later that summer. His biographers, Kai Bird and Martin Sherwin, summarized the speech as follows:

> It was an extraordinarily perceptive and brazen speech. Again and again, Oppenheimer observed that he was barred from speaking of the essential facts—and then like a Brahmin priest endowed with special knowledge, he proceeded to reveal the most fundamental secret of all—that no country could expect in any meaningful sense to win an atomic war... [I]n a chilling turn of phrase that startled everyone who heard it, Oppenheimer quietly added, "We may be likened to two scorpions in a bottle, each capable of killing the other, but only at the risk of his own life."[17]

Oppenheimer went on to argue that the world's only hope of survival lay in "candor"—the international sharing of nuclear stockpile and testing information so that no one nation could develop a decisive advantage.

Whether due to coincidence or prescience, Oppenheimer's "candor" speech appeared in *Foreign Affairs* a month before Operation Joe, the first Soviet test of a hydrogen bomb. Oppenheimer's proposals consequently gained some purchase within the Eisenhower administration.[18] The president gave both a domestic press conference and a speech to the United Nations that year invoking "candor," though in a much weaker sense than Oppenheimer had intended. Nevertheless, Strauss hit the roof. Not only was the proposed policy anathema to the hawkish admiral, but more importantly, he felt Oppenheimer had overreached his authority and embarrassed Strauss and the AEC by issuing such a potent policy statement unilaterally.[19]

Strauss decided to take Oppenheimer out of politics.[20] He had aides dredge up derogatory FBI information on Oppenheimer—much of it obtained through illegal wiretaps—that linked him to American Communists in the 1930s and early 1940s. This information was old news, having been thoroughly vetted by Manhattan Project security supervisors during Oppenheimer's clearance proceedings in 1942. But on the basis of it, Strauss managed to convince Senator Joseph McCarthy, who had Oppenheimer high on his list of suspected "reds," to leave his case to the AEC. Strauss could have just let Oppenheimer's clearance expire as scheduled in June 1954; as Isidor Rabi pointed out later during the hearings, he also could have simply elected not to use Oppenheimer as a consultant, since that was the only capacity in which he was serving the AEC at the time. But Strauss wanted Oppenheimer humiliated.[21]

On December 3, Strauss presented a brief of the evidence against Oppenheimer—much of it dusty, some outright fabricated—to President Eisenhower, who was apparently shaken enough at its allegations to order a "blank wall" placed between Oppenheimer and national secrets;[22] Oppenheimer's security clearance was suspended pending review by the AEC's Personnel Security Board (commonly referred to as the Gray Board for its chair, Gordon Gray).

Herbert Block, the gifted *Washington Post* editorial cartoonist, depicted Eisenhower's blank wall literally—with Eisenhower, Strauss, and Uncle Sam scratching their heads on one side, and Oppenheimer sitting on the other side with the atom in his hands (figure 6). The caption read, "Who's Being Walled Off From What?" From the standpoint of ethos, the cartoon wall stood for the slash in the is/ought model, the putative boundary between science and policy, disciplinary ethos and prophetic ethos. The untenability of the model had never been more apparent.

The blank wall also represented national security. This, indeed, would become the central topos in the Gray Board hearings in April 1954. In advance of those hearings, Oppenheimer responded to Strauss's charges with an

FIGURE 6 "Who's Being Walled Off From What?" The wall is a reference to Eisenhower's directive in December 1953 that a "blank wall" be placed between Oppenheimer and all national secrets pending his security review by the Gray Board. President Dwight D. Eisenhower and Admiral Lewis Strauss, the chairman of the Atomic Energy Commission, are depicted in frantic consultation with Uncle Sam about how to get access to the secrets of the atom, which Oppenheimer holds in his hand. Rhetorically speaking, I argue the wall signifies the "is/ought" divide in American science-policy integration.

Source: A 1954 Herblock Cartoon, copyright by The Herb Block Foundation.

autobiography, widely reprinted, in which he told the story of the formation of his character in an effort to demonstrate his loyalty and put in perspective his contacts with Communists in the 1930s. These two documents, the charges and the autobiography, formed the rhetorical grounds, first, for the hearings, the transcript of which runs to nearly 1,000 pages; and second, for the findings report, which presented and justified the Board's verdict (recommending the AEC revoke Oppenheimer's clearance). Only one of the three board members, Dr. Ward Evans, dissented from the majority opinion and contributed a spirited defense of Oppenheimer's character and loyalty, denouncing the proceedings as a "black mark on the escutcheon of our country."[23]

The hearings are now generally considered to have been a kangaroo court, with Strauss pulling strings behind the scenes to guarantee Oppenheimer's defeat:[24] his tactics ranged from wiretapping Oppenheimer's lawyer's office to refusing to let the defense see the evidence the Gray Board had in front of them.[25] Several of Oppenheimer's friends, including Einstein, had advised Oppenheimer not to testify at all. But he insisted that he could not let the charges stand because "[i]f I were thus unworthy I could hardly have served our country as I have tried...or have spoken, as on more than one occasion I have found myself speaking, in the name of our science and our country."[26]

The purpose of the hearings was to determine if Oppenheimer was enough of a security risk to have his clearance revoked. Security for these purposes was defined by the language of Executive Order 10450, which set out the Eisenhower administration's new restrictions on federal employees and which required federal agencies, such as the AEC, to enforce them: this was the pretext under which Strauss, as AEC chair, attacked Oppenheimer; Strauss was only doing his duty by Order 10450.

Order 10450 contained predictable language about the giving of American secrets to foreign agents being "[in]consistent with the interests of the national security." But section 8(a) also contained vaguer and more troubling clauses defining "behavior" that constituted a "security risk." This behavior included "but shall not be limited to...any behavior, activities, or associations which tend to show that the individual is not reliable or trustworthy," and "immoral, or notoriously disgraceful conduct, habitual use of intoxicants to excess, drug addiction, sexual perversion." Here we can see how security is really an attempt to perpetuate political certainty. Order 10450 was not just trying to protect information, it was trying to enforce the performance of "American" Judeo-Christian mores. Via this expansion of the notion of "security," the Gray Board suddenly had jurisdiction over Oppenheimer's moral life.

The security charges the Gray Board had in front of them fell under two headings: first, that Oppenheimer's prior and ongoing friendships with Communists made him a security risk; second, that his lack of "enthusiasm" for the hydrogen bomb project had delayed the project and potentially divided his loyalties as a federal consultant.[27]

As I have already indicated, the Gray Board got nowhere with the first set of charges. The discussion of the second category, however, is of particular interest to our analysis of scientific-prophetic ethos. These charges implied that if a scientist made policy evaluations that were in line with the agendas of dominant federal constituencies, such as the military or the Executive, this was all right; but if the policy evaluations ran counter to current executive or military policy, such statements nullified the scientist's status as a loyal citizen. This is not just my reading of the H-bomb charges; it was also the reading of many of the nearly two dozen scientists who testified on Oppenheimer's behalf in the Gray Board hearings (including James Conant, Hans Bethe, Vannevar Bush,

John McCloy, and Isidor Rabi),[28] of 214 Los Alamos scientists who signed a protest after the hearings were over,[29] and of the authors of many clippings archived in Oppenheimer's papers.[30] Vannevar Bush was particularly forceful in criticizing the Gray Board for trying Oppenheimer's political opinions. He famously thundered, "If you want to try that case, then you can try me. I have expressed strong opinions many times, and I intend to do so. They have been unpopular opinions at times. When a man is pilloried for doing that, this country is a severe state."[31]

Clearly at the heart of the H-bomb charges was the proper ethos of science advisers. Both Oppenheimer's testimony and the Gray Board's response to it reveal simultaneously (a) confirmation of Oppenheimer's performance of the prophetic meme; and (b) attempts to constrain this performance with the is/ought model.

Oppenheimer reliably performed many aspects of the prophetic meme in his Gray Board autobiography and testimony, just as he did in the executive and public texts we reviewed earlier. He presented himself in his autobiography as a young man so unworldly and devoted to his studies that he didn't even read the newspaper. He recounted the ascetic rigors of Los Alamos, concluding that he and his colleagues had borne the hardships because they were a community "inspired by a high sense of mission, of duty and of destiny, coherent, dedicated and remarkably selfless." He performed humility in claiming that the success of the project was "the opposite of a one-man show." He spoke of a calling, a "feeling of deep responsibility" that he should help guide atomic policy even though he wanted to go back to his quiet "academic life."[32] And throughout his testimony to the Gray Board, Oppenheimer persisted in his habit of invoking secrecy, knowing the Board had been cleared to look at many classified documents but not the Top Secret atomic information he knew.

He also injected some prophetic stylings into his testimony—aphorisms, moral judgments, and predictions. A sample of his aphoristic responses: when asked to explain why he opposed a "crash" (accelerated) program to build an H-bomb when many thought the Russians would soon have one, Oppenheimer replied he had believed that "[s]ometimes the only answer for fear lay in courage."[33] And in a rare and bleak stab of quantum humor, when Chairman Gray remarked on Oppenheimer's return to the room during Rabi's testimony, "You are back now, Dr. Oppenheimer," Oppenheimer quipped, "This is one of the few things I am really sure of."[34]

Although this was the very thing he was on "trial" for, Oppenheimer persisted in making value judgments and predictions about atomic policy during the hearings, as seen in the following volley with Gray Board counsel Roger

Robb over Truman's decision to support the crash program against the GAC's advice:

> Q. [Robb]. Did you subsequent to the President's decision in January 1950 ever express any opposition to the production of the hydrogen bomb on moral grounds?
>
> A. [Oppenheimer]. I would think that I could very well have said this is a dreadful weapon, or something like that. I have no specific recollection and would prefer it, if you would ask me or remind me of the context or conversation that you have in mind.
>
> Q. Why do you think you could very well have said that?
>
> A. Because I have always thought it was a dreadful weapon. Even [though] from a technical point of view it was a sweet and lovely and beautiful job, I have still thought it was a dreadful weapon.[35]

When pressed further on his views about the decision, Oppenheimer responded that "there was danger in the fact that such decisions had to be taken secretly, not because the people who took the decisions were not wise, but because the very need, the very absence of criticism and discussion tended to corrode the decision making process."[36]

The Gray Board, and even the overtly hostile Robb (who described Oppenheimer as "just a brain... as cold as a fish" with "the iciest pair of blue eyes I ever saw")[37] recognized many of these prophetic performances, thus helping to constitute Oppenheimer's ethos as a scientist-prophet. Isidor Rabi nodded to this constitution when he reminded the Gray Board: "That is what you are really doing here. You are writing a man's life."[38]

First, the Board acknowledged Oppenheimer's standing as an eminent physicist. Although they made a point of addressing both Colonel John Lansdale, the security officer for Los Alamos during the war, and Ambassador George Kennan as "Mr.," they consistently addressed Oppenheimer, and in fact all the scientists who testified, as "Doctor." Robb further reinforced the Board's appreciation of scientists' privileged access to knowledge when he refused to accept the claim of David Lilienthal, former AEC chair, that he had developed his opinion about Russian H-bomb capabilities on his own. Robb ridiculed him: who told him what the Russian situation was? Surely he did not get his information from other lawyers, from "Mr. Garrison [Oppenheimer's counsel] or myself." Lilienthal clearly must have "check[ed] with the scientists."[39] In Robb's world, there was no other portal to that kind of knowledge.

The Board ascribed to Oppenheimer's exceptionality in other ways as well. While they disputed his final conclusions about Oppenheimer's trustworthiness, they did not dispute Col. Lansdale's preferential security treatment of Oppenheimer due to his "importance" to the atom bomb effort.[40] And they "very much enjoyed the discussion" in which George Kennan compared Oppenheimer to the prophetic Saint Francis:

> I sometime think that the higher types of knowledge and wisdom do not often come without very considerable anguish and often a very considerable road of error. I think the church has known that. Had the church applied to St. Francis the criteria relating solely to his youth, it would not have been [possible] for him to be what he was later. In other words, I think very often it is in public life as it is in the life of the spirit; it is only the great sinners who become the great saints and in the life of the Government, there can be applied the analogy.[41]

But just as they acknowledged Oppenheimer's prophetic ethos, the Gray Board also made it clear that they thought he had overstepped his proper role. They tried to show that Oppenheimer had violated the is/ought model in two ways. First, they tried to establish that as GAC chairman, Oppenheimer should have restricted himself to giving only technical advice. In a humiliating gesture, Robb even made Oppenheimer repeat that phrase from the Atomic Energy Act (1946) charter of the GAC twice.[42] The Board concluded from this charter that Oppenheimer had no business making value judgments or policy recommendations in his capacity as GAC chair; they hammered him on these rhetorical points for hours.[43] But this attack fizzled after witnesses made it clear that a more progressive ethical model had operated under Lilienthal's chairmanship of the AEC, during which time the commission had often charged the GAC with policy questions.[44]

The Board then tried a new tack. They pursued the charges in Nichols's letter stating that Oppenheimer had tried to block Truman's crash program directive by inveighing against it in public media and by discouraging colleagues from going to work with Teller and Ernest Lawrence in their H-bomb lab at Lawrence Livermore in Berkeley. Unable to get any scientists to swear that Oppenheimer had discouraged them from H-bomb work after 1950, they brought Teller to the stand.

Teller was deeply resentful of Oppenheimer's fame and furious that Oppenheimer had advised against the crash program, which hinged on an H-bomb prototype Teller had sketched;[45] Oppenheimer had joked it would take an "ox-cart" to deliver the unwieldy design to its target.[46] Teller read this move not as a principled moral or technical stance but rather as a territorial unwillingness to share the fruits of political celebrity that came with atomic weapons development. In another of his outrageous abuses of legal protocol, Strauss had Teller rehearse his testimony with an assistant before the hearings, during which meeting Teller said he wanted to "deepen the charges" concerning resistance to the H-bomb. The assistant replied that it sounded as if Teller wanted to see Oppenheimer "unfrocked in his own church"; Teller concurred with the phrasing.[47]

When asked under oath if he felt Oppenheimer's anti-H-bomb sentiments put him at odds with the interests of national security, Teller answered: "His actions frankly appeared to me confused and complicated. To this extent I feel that I would like to see the vital interests of this country in hands which I understand better, and therefore trust more."[48] Later, Teller added, "If it is a question of wisdom and judgment, as demonstrated by action since 1945, then I would say one would be wiser not to grant clearance."[49] Teller's self-serving testimony ironically confirmed the majority reading of the H-bomb charges: weapons scientists were free to operate according to the progressive model as long as their recommendations sailed with the prevailing policy winds in the War Department (personified in this case by the hawks Admiral Strauss, General Nichols, and David Griggs, chief Air Force scientist); if scientists' advice tacked off, however, they were disloyal citizens.

The Gray Board's *Findings and Recommendations* report, submitted to the AEC on May 27, concluded, "We find his conduct in the hydrogen bomb program sufficiently disturbing as to raise a doubt as to whether his future participation, if characterized by the same attitudes in a Government program relating to the national defense, would be clearly consistent with the best interests of this country."[50] The last phrase is quoted from Executive Order 10450, and it enforces the impression that the Gray Board, too, was defining security as the future chances of Oppenheimer performing their covenant values. The Board justified the finding as follows:

> We must make it clear that we do not question Dr. Oppenheimer's right to the opinions he held with respect to the development of this weapon. They were shared by other competent and devoted individuals, both in and out of Government. We are willing to assume that they were motivated by deep moral conviction. We are concerned, however, that he may have *departed his role as scientific adviser* to exercise highly persuasive influence in matters in which his convictions were not necessarily a reflection of technical judgment, and also *not necessarily*

> *related to the protection of the strongest offensive military interests of the country.*(emphasis mine)[51]

Here the application of the is/ought model is very clear, as is Strauss's reactionary interpretation of security, in the "strongest military offensive interests" language. The decision ignores that (a) it was clear from testimony that Oppenheimer was directly asked to give policy recommendations, and (b) it is unrealistic to expect scientists not to articulate values and/or emotions when making policy recommendations, since *values must be attached to technical states of affairs to motivate political action.* In his dissenting opinion, Ward Evans homed in on the reaction of his fellow scientists to such unrealistic expectations for their behavior when serving the government as science advisors: "[Oppenheimer's] witnesses are a considerable segment of the scientific backbone of our nation and they endorse him. I am worried about the effect an improper decision may have on the scientific development in our country."[52]

The AEC reviewed the Gray Board report, and then voted four to one to strip Oppenheimer's clearance, with scientist Henry Ward Smyth dissenting. Interestingly, in its report, perhaps keying in on the furious tenor of the scientific testimony and Evans's concerns about the chilling effect of the proceedings on young scientists, the AEC claimed the H-bomb charges had nothing whatsoever to do with its decision. It claimed it deemed Oppenheimer a security risk due to his dusty "associations" with Communists. But in going through the motions of rejecting the H-bomb charges, Commissioner Thomas Murray ironically reaffirmed the scientific community's majority reading of them. Murray argued that when serving under the auspices of a security clearance, scientists were bound to carry out federal policies without "disobeying" them. Naturally, Murray argued, federally cleared scientists were free to have any of a number of "feelings" about federal policy; they could even "voice... moral doubts." What followed was a prize case of doublespeak:

> The citizen's duty remains always that of reasonable service, just as the citizen's right remains always that of free opinion. There is no requirement, inherent in the idea of civic duty, that would oblige a man to show enthusiasm for particular governmental policies, or to use his influence in their favor, against his own convictions; *just as there is no permission, inherent in the idea of intellectual freedom, that would allow a man to block established governmental policies against the considered judgment of their responsible authors* (emphasis added).[53]

In short, under Murray's paradigm, science advisers were the administration's policy robots. Since the Gray Board had been unable to establish evidence of any direct actions against national security—leaking secrets or dissuading scientists from following Truman's orders—and the Communist associations were tissue thin at best, the AEC report left the strong impression that, in spite of their protests to the contrary, Oppenheimer was indeed being condemned for his public statements against the H-bomb. Somehow, these had "blocked" federal policy.

This impression is substantially confirmed by the public reception of Oppenheimer's ethos as performed in the Gray Board hearings.[54] The Oppenheimer papers contain six folders full of clippings of reviews and op-ed pieces on the hearings from papers all over the world; they represent only a fraction of the media coverage of the event. The majority of authors, both domestic and foreign, supported Oppenheimer. This response is probably representative of the overall public evaluation of the Oppenheimer case even though the sample in his papers is undoubtedly skewed somewhat in his favor.[55] But there are certainly enough critical arguments contained in the clippings that the overarching topoi of the public reaction can be summarized as follows.

The group who supported Oppenheimer's innocence protested that his civil rights were being trampled, that he was subject to an "inquisition." Several writers suspected he was the sacrificial victim of an internal political feud in the AEC. A few writers even compared Oppenheimer to the prophetic character of Prometheus, who was punished for trying to share the gods' secrets with mortals. This group repeated the concern expressed by the scientists at the Gray Board hearings that disagreeing with current military policy constituted "un-American" behavior. They argued that Eisenhower's "blank wall" made no sense because Oppenheimer already knew all of America's nuclear secrets. Finally, these writers worried about recruitment of young and foreign scientists and called for reform of the security code (particularly Executive Order 10450).

This cluster of arguments was repeatedly referred to as "liberal" by writers supportive of the Gray Board's verdict. The writers in this group argued that desperate times called for desperate measures; hence, national security must be ranked above personal liberty. Some of these writers leaned on Senator McCarthy's insinuation in a television address in December 1953 that Oppenheimer had stymied the H-bomb.[56] A few writers picked up on Oppenheimer's statement in his autobiography, which was widely reprinted along with transcript excerpts, that before World War II he had been so absorbed in his scientific and philosophical studies that he had not even read newspapers; writers attributed his eventual downfall to this cloistered lifestyle, which left the scientist "over-educated" and "lacking common sense," in their opinion.

In short, the public reception evidence zeros in on the proper ethos of a scientist involved in government: we see again the conflict in the progressive and is/ought models, with one group arguing Oppenheimer was only being punished for what he had been asked to do and the other group arguing that he was culpable for not observing the traditional limits on his public behavior.

The Fermi Prize, 1963

After the humiliation of the Gray Board and AEC hearings, Oppenheimer went into seclusion in the Caribbean with his family for several months. When he came back, it was to a strange ethical landscape. He found a few ties cut: his files contain letters rescinding two invitations to speak or teach at universities.[57] But he retained his position at the IAS in spite of Strauss's attempts to get him fired, and his files are full of invitations to speak elsewhere—especially in Europe and at conferences examining atomic politics. This was how Oppenheimer occupied himself for the remainder of his career: no longer constrained by the government (beyond the surveillance that would follow him for the rest of his life), he now fully embraced the role of kairotic prophet, engaging polities at home and abroad in dialogue on the nuclear future and the imperative to build a common culture. One example of Oppenheimer's prophetic language from this time comes from a speech he gave for the Columbia University Bicentennial:

> In the arts and in the sciences it would be good to be a prophet. It would be a delight to know this future. I had thought for a while of my own field, of physics, and of those fields nearest to it in the natural sciences. It would not be too hard to outline the questions that natural scientists today are asking themselves, trying to answer. What, we ask in physics, is matter? What is it made of? How does it behave when it is more and more violently atomized, when we try to pound out of the stuff around us the ingredients which only violence creates and makes manifest?...All history teaches us that these questions, that we think are the pressing ones, will be transmuted before they are answered, that they will be replaced by others, and that the very process of discovery will shatter the concepts that we today use to describe our puzzlement.[58]

Another example comes from an episode of the radio program *The World We Live In*. After expressing his hopes for better "translation" between scientific and public spheres, Oppenheimer stressed that "[a]bove all, it is good that in this translation the vision of order, wonder and beauty in the world should not be lost. This vision is ennobling for us all; this is a reason to learn about science for the sake of our humanity."[59]

When the Kennedy administration came into power in 1963, it marked a sea change in the way government dealt with science. The Democratic Party had cultivated several scientific advisory boards: Rachel Carson sat on one, Leo Szilard on another. Kennedy immediately reformed the defunct President's Science Advisory Committee (PSAC) and appointed James Killian its chair and his Special Assistant on Science and Technology. Killian also liaised with the Office of Science and Technology, which was responsible for coordinating advisory efforts throughout the Executive. The "liberal" camp in the reception of Oppenheimer's trial had come to power with progressive tendencies that made it more comfortable with integrating scientific advice in policy making. NASA's budget skyrocketed with the approval of the Apollo missions, the president ordered a thorough review of federal spraying programs in the wake of the publication of Carson's *Silent Spring*, and Kennedy made it known that he wished to give the relatively new Fermi Award for atomic achievement to Oppenheimer. The public responses contained in Oppenheimer's papers, both pro and con, make it clear that the medal was widely viewed as an indirect federal apology for the Gray Board hearings.

The new AEC chairman, Glenn Seaborg, and other powerful friends quietly sounded Oppenheimer out about whether he wished to go through a series of new hearings to have his clearance reinstated. His response: "Not on your life."[60] The Gray Board hearings had aged him prematurely, and he was increasingly weakened by the lung cancer that would end his life in 1967 at age 62.

Oppenheimer was presented with the Fermi Award on December 2 of that year by Lyndon B. Johnson, who had assumed the presidency just ten days prior following Kennedy's assassination. In his very short acceptance speech at the White House ceremony, Oppenheimer looked down at the medal for a few moments before turning to President Johnson and saying, "I think it is just possible, Mr. President, that it has taken some charity and some courage for you to make this award today. That would seem to me a good augury for all our futures." He paused and then added, "These words I wrote down almost a fortnight ago. In a somber time, I gratefully and gladly speak them to you." In casting his speech as a prediction, Oppenheimer made a final performance of prophetic ethos in the halls of government.

Two years after Oppenheimer's death, political journalist Phillip M. Stern published an intensively researched denouncement of the Gray Board hearings called *Security on Trial*. His reflections on the definition and prosecution of security in the United States helpfully clarify and extend the way I have formulated scientific-prophetic ethos to this point. Stern's central argument was that the most extreme actions the United States took to secure itself ironically made the country less secure: draconian clearance protocols scared off the best and brightest scientists from working for the government, and scientists exiled during the McCarthy era went to work for foreign nations.[61] But while security was commonly configured as being about things—objects and information—Stern argued that it was really about values, about certainty. He showed that as the loyalty-security protocols intensified under Order 10450, the clearance questions put to candidate government scientists became oriented less toward the objects of security and more toward covenant values—at least the covenant values espoused by the conservative Nixon administration: "What do you think of female chastity? Is it proper to mix white and Negro blood plasma? There is a suspicion in the record that you are in sympathy with the underprivileged. Is that true?"[62]

The shock of these questions to the twenty-first century ear is a powerful reminder that security is really about maintaining covenant values through crisis. In this way, a drama ostensibly about the protection of nuclear secrets becomes a drama about race and class relations. In Stern's succinct formulation, America's "fear about the outside world" was "actually a fear about itself."[63] Under this interpretation, Oppenheimer as cultic scientist-prophet was "unfrocked" not because he had compromised America's safety but because he had disrupted its certainty: his calls to Americans to evaluate the nuclear crisis in terms of internationalism and peace had interfered with the Eisenhower administration's attempts to cast the crisis in terms of nationalism and militarism. But not all prophets were so easily disciplined. In the next chapter, we shall see how Rachel Carson successfully projected the voices of marginalized Others into the sanctuary of national science policy.

7 RACHEL CARSON

KAIROTIC PROPHET

In many ways 1963 was a banner year for American science. The budget for the historic Apollo moon missions was approved by Congress. J. Robert Oppenheimer was given the Fermi Medal for a lifetime of service in nuclear energy research. And in May, the PSAC issued a long-awaited report on pesticide use that endorsed most of the recommendations from a controversial book written by an independent biological researcher: *Silent Spring* (1962).

Although Carson and Oppenheimer shared the laurels bestowed by the new, progressive administration in 1963, and both were born and died (of cancer) within three years of each other, their biographies stand in otherwise sharp contrast. He was an upper-class Jew by birth, a publisher of groundbreaking physics, director of a national laboratory, adviser to presidents; she was a working class Protestant with a terminal master of science in biology who toiled for years writing reports for the US Fish & Wildlife Service (USFWS) before becoming a freelance writer. Notwithstanding these differences, Oppenheimer and Carson both leveraged prophetic ethos to push for more public dialogue on key scientific issues. Just as Oppenheimer exploited his privileged access to the invisible workings of the atomic world to bolster his ethos, Carson used her micro- and telescopic visions of ecosystems to dramatize her prophetic claims about human impacts on the environment. She galvanized political action via a rhetorical strategy that combined prophetic battle with apocalyptic visualizations of risk.

Rachel Carson was born on a farm in Springdale, Pennsylvania, in 1907. A precocious writer, she managed to secure a scholarship to the Pennsylvania College for Women. Initially, she loved writing and wanted to pursue a career in English literature, but she found herself mesmerized by the lectures of Mary Scott Skinker, a charismatic biology professor. Carson's journals from this time reveal her searching for a vocation, a "vision splendid," in her words.[1] Torn between her loves for science and literature, she became anxious and depressed about her future career prospects: in the 1920s such prospects were dramatically limited for women.

In 1928 Carson had a sort of conversion experience reminiscent of Robert Boyle's. She was in her attic dorm room reading "Locksley Hall" by Tennyson for a class assignment. A violent storm rattled the windows in her room and lashed the panes with rain as Carson read the line, "For the mighty wind arises, roaring seaward, and I go." It suddenly came to her that her "destiny," her "vision splendid" lay in writing about the sea. It was the perfect way to combine her passions for science and literature.[2]

Carson graduated in 1929 and entered the marine biology program at Johns Hopkins University as one of only two women in a class of 70. By the time she got her MS in zoology in 1932, she had logged significant hours in the lab at Woods Hole Oceanographic Institute in Massachusetts and on field trips to places such as the Florida Everglades. She landed her first job as a technician for the USFWS, but when her supervisor noticed her writing skills, she was quickly promoted in that direction. By 1949, she was the editor in chief at the USFWS publications bureau. At the same time, she was working on her own essays, reflecting on her marine experiences and what she was learning about the budding science of ecology. She published some of these essays in the *Atlantic* and the *Baltimore Sun*, and Simon and Schuster offered her a book contract. *Under the Sea-Wind* was published in 1941, and Oxford University Press released *The Sea around Us* in 1951; it was serialized by the *New Yorker* and spent 86 weeks on the *New York Times* best-sellers list. After this success, Carson was able to retire from the USFWS and devote her career to writing popular works on ecology. These sold well enough for her to maintain her house in Maryland and take on a summer home on Southport Island in Maine.

Although she enjoyed a loyal following and accolades for her elegant rhetorical handling of ecology, it was pesticides that would make her a household name. She had become aware of the wide-ranging effects of pesticides during her work for the USFWS. She was troubled by the effects on fish and waterfowl of dichlorodiphenyltrichloroethane (DDT), with which wetlands were being carpet bombed to control mosquitoes. Then, in 1957 and 1958, several events converged that caused her to take up her pen against pesticides: a disastrous United States Department of Agriculture (USDA) attempt to eradicate fire ants in the South using dieldrin, a broadly toxic chlorinated hydrocarbon related to DDT that did stunning collateral damage to livestock and wildlife;

a high-profile lawsuit on Long Island prosecuted by conservation friends of Carson's against the state government after a DDT spraying campaign against gypsy moths illegally canvassed commercial dairy farms, killing livestock and sickening humans; and, a deluge of letters from fans complaining about related campaigns against the elm bark beetle and the mosquito. Carson initially wrote to E. B. White, the *New Yorker*'s nature essayist at that time, to suggest he cover the Long Island trial. White demurred due to his schedule and offered her the story instead. That was how Carson embarked on the series of *New Yorker* essays that would be collected in *Silent Spring* in 1962.

The work on the book was difficult and tedious, in part because some academics and industry scientists were resistant to sharing information with an independent researcher,[3] in part because she was caring for her orphaned nephew at the time, and in part because she fell ill with a rapidly metastasizing breast cancer. Her contract had stipulated delivery of the manuscript by September 1960; she would not deliver it for two more years. But she drove herself to complete the book, working from bed with an assistant when she was too exhausted by chemotherapy to sit at her desk. When her friend Dorothy Freeman urged her to put her health first, Carson replied, "Knowing what I do, there would be no future peace for me if I kept silent... it is, in the deepest sense, a privilege as well as a duty to have the opportunity to speak out—to many thousands of people—on something so important."[4] Pesticides were for Carson a metonym standing in for the myriad invisible wounds humans were ignorantly inflicting on ecosystems. She felt a calling to wake Americans up to the effects of their unconscious choices.[5]

Silent Spring

When her articles appeared in the *New Yorker* in six installments beginning in June 1962, they made an instant sensation. The *New Yorker* received more mail than it ever had for a single author; the majority of letters praised Carson for her "public service" in calling attention to the pesticide issue.[6] The forthcoming book was immediately preordered for the Book of the Month Club and was scheduled for additional print runs in anticipation of demand; they all sold out. Houghton Mifflin received an equally stunning volume of mail when the book came out in September, as did Carson herself, as did local newspapers. The letters were overwhelmingly supportive, coming roughly equally from men and women, with a slight edge for women in regional newspapers. Many letters came from concerned physicians.[7]

While my primary interest in this chapter is the federal reception of Carson's arguments in *Silent Spring*, in this case it is clear to historians that the government only took up the issue of pesticides and called Carson to testify because of the enormous public response to her book.[8] Accordingly, how the government received Carson's prophetic ethos depends heavily on its constitution by popular media.

A great deal has been written about the rhetoric of *Silent Spring*, most of it aiming to find out how Carson's strategies managed to convince so many Americans to write their congresspeople, throw out their DDT, or otherwise take civic action. The problem with most of these analyses is that from our present-day perspective, they draw an explanatory line from the banning of DDT in 1972 back through the formation of the EPA in 1970 and the congressional hearings in 1963 to particular things Carson wrote. But if we look at public arguments about pesticides in 1962 and 1963 more closely, we find many more agents involved in the drama. Although the book undoubtedly had a huge civic impact, it was just one spasm in a crisis of awareness.[9] Carson herself was quick to admit that her book came at just the right time. The Great Cranberry Scare of 1959 had sensitized Americans to pesticide residues on their Thanksgiving tables. And just before *Silent Spring* came out, Murray Bookchin had published a book on pesticide risks called *Our Synthetic Environment* (he did so under the pseudonym Lewis Herber for fear of reprisals from the chemical industries).

In addition to these in-progress conversations about the risks of pesticides, Carson leveraged the growing public fear of nuclear fallout. People were horrified by news of the Lucky Dragon incident, in which fallout from an American H-bomb test in the South Pacific sickened the crew of a Japanese fishing boat. More terrifyingly for American mothers, doctors had recently discovered strontium-90, a radioactive by-product of atomic fission, in cow's milk and in all baby teeth tested. Carson's opening fable in *Silent Spring* depicted pesticides falling in a silent, white powder, both mimicking the Lucky Dragon fallout and evoking the whiteness and ubiquity of milk; elsewhere in the book, she made the comparison more explicit.[10]

Carson's book thus benefited, both deliberately and accidentally, from these fears swirling in the kairos surrounding its publication. However, the seed crystal that coalesced these arguments into an exigence for the government was the CBS Reports television program *The Silent Spring of Rachel Carson*, which aired on April 3, 1963. Given that many people who argued the merits of *Silent Spring* never read the book but instead responded to media presentations of its arguments,[11] we must take into consideration the broadcast segment of the *Silent Spring* drama. It is widely recognized that the outpouring of letters to federal agencies and legislators in the wake of the program goaded the PSAC into releasing its long overdue report on pesticides, for which Carson was interviewed in January 1963; similarly, the public response prompted the Ribicoff congressional hearings on pesticide use, at which Carson also testified in June 1963.[12] We can use the television broadcast, therefore, as a central lens through which to examine Carson's prophetic ethos as articulated via her privileged vision of the ecosystem, her prophetic battle with representatives of American chemical industries, and her dramatizations of abstract risks.[13] Then, we will be in a better position to appreciate the federal reception of Carson's ethos.

Speaking for the Silenced: Carson's Privileged Micro-/Telescopic Vision

When *Silent Spring* was first published, the CBS anchorman Eric Sevareid had published a response in the *Los Angeles Times* that both picked up on the apocalyptic connections between fallout and pesticides and gestured toward Carson's prophetic ethos: "It is quite wrong for us to assume that in atomic war lies the only danger of 'setting back civilization' a thousand years. . . . The new religion of the scientist-philosopher, like the old time religion, invokes the sanction of hell-fire and damnation—but with proof."[14] Sevareid went on to direct the influential CBS Reports program covering Carson's book. As became apparent, he was solidly on her side, although she was worried during the production that he would not cast her or her argument in a favorable light.[15]

Sevareid conducted the bulk of his interviews with Carson in her living room in Silver Springs. But key action sequences show her at her microscope or emerging from the Maine woods with a pair of binoculars around her neck. Images of her with these magnifying instruments are nearly ubiquitous in media coverage of her book and career. One such image in *Life* magazine, in which she is leading a group of Audubon Society members on a field trip near her home, captures everyone peering skyward through binoculars at some distant wonder. Carson herself jokingly referred to this photo as "the Second Coming."[16] But it was precisely her special ability to see and describe systems outside the limits of normal human vision that enabled her to engage Americans in consequential dialogue about the environment.

Carson synthesized microscopic and telescopic views to demonstrate the systemic damage wrought by pesticides. She began by arguing that the structure of the cosmos could be grasped via the study of microcosms. Chapter 13, "Through a Narrow Window," opened with a quote from a scholar of ocular pigmentation who justified his specialization by comparing it to a narrow window. True, the window would offer little perspective to the casual or distant observer; however, "as one comes closer the view grows wider and wider, until finally through the same narrow window one is looking at the universe." Carson concluded:

> So it is that only when we bring our focus to bear, first on the individual cells of the body, then on the minute structures within the cells, and finally on the ultimate reactions of molecules within these structures—only when we do this can we comprehend the most serious and far-reaching effects of the haphazard introduction of foreign chemicals into our internal environment.[17]

The beautiful engravings by Louis and Lois Darling of exploded cells and soil microbes that headed this and other chapters of *Silent Spring* reinforced Carson's devotion to the elegant order of invisible natural systems.[18]

The CBS program reinforced Carson's argument by devoting almost a third of its airtime to her reading the sections of *Silent Spring* in which she argued seemingly disparate life forms into configuration. Here is one of those sections:

> We poison the caddis flies in a stream and the salmon runs dwindle and die. We poison the gnats in a lake, and the poison travels from link to link of the food chain, and soon the birds of the lake margins become its victims. We spray our elms and the following springs are silent of robin song, not because we sprayed the robins directly, but because the poison traveled step by step, through the now familiar elm leave-earthworm-robin cycle.[19]

The effects of DDT were "biomagnified" in their cascading effects on the whole ecosystem. Furthermore, they were invisible to everyone but scientists. During the CBS broadcast, Carson also read the following chilling sentence from chapter 11, "Beyond the Dream of the Borgias:" "In river or lake or reservoir, or for that matter in the glass of water served at your dinner table, are mingled chemicals that no responsible chemist would think of combining in his laboratory." Here she both brought the dinner table into focus as part of the "web of life" and demonstrated her privileged microscopic vision of the toxins being "served" there.

The "Fable for Tomorrow" opening the book was just such a synthesis of telescopic and microscopic vision. It amounted to a jeremiad on America's future should it refuse to change course. The fable described an idyllic small town "in the heart of America where all life seemed to live in harmony."[20]

Following a dusting with pesticides, however, spring brought no new wildflowers, no birds, no bees, just silence: "No witchcraft, no enemy action had silenced the rebirth of new life in this stricken world. The people had done it themselves."[21] The town was a projection of a dystopian future based on scientific intervention, just as surely as Bacon's *New Atlantis* projected a utopia on the same basis.

Carson's most impressive telescopic feat was combining the unheard voices of thousands of citizens, mostly women, and amplifying them into a resounding indictment of the collusion of industry, government, and academic scientists. Carson frequently mentioned the hundreds of letters she received from readers of her books reporting their hand tremors, dead pet cats, leukemia, and birdbaths full of dead robins.[22] She even publicly attributed her decision to write *Silent Spring* to one such letter she had received from Olga Huckins asking Carson where all the songbirds who normally visited her birdfeeder had gone and pleading with her "if something couldn't be done."[23] Carson clarified this prophetic vision of a suffering, marginalized people in a speech to the National Parks Association in October 1962:

> I came to realize that scattered throughout the country were thousands of people who were concerned—who were trying, as individuals or as small groups, to do what they could, in the face of great odds. Now, simply because I happen to have brought together the basic facts—because I have written a book that seems to be serving as a rallying point for an awakened public—both the strength and the needs of these people are flowing to me in a vast and wonderful way.[24]

Just as the individual dishes of a radio telescope array combine their signals to produce a coherent vision of a dying star, Carson's dialogue with silenced and ignored American housewives, pensioners, and pediatricians enabled her to synthesize a vision of the whole damaged ecosystem—and of the dysfunctional political system. In *Silent Spring*, in the CBS program, and in her congressional testimony, she mounted an awe-inspiring prophetic micro-/telescopy that confronted the government with the coherent presence of the Other—both human and natural.

False Prophets: Carson and the Agrichemical Industry

The basic structure of the CBS program proceeded like a classical debate. Carson's questions about pesticides were put to agricultural chemists and federal administrators, and in turn, Carson answered their charges.

Priscilla Coit Murphy reminds us that this is often the way that news media concentrate a diffuse *issue* like pesticide use into an *event*, the natural object of journalism.[25]

The result of this rhetorical reduction was a trial of ethos—Carson versus the chemists. According to the rules of media debates, there could be only one winner. That meant that in the CBS report and more generally in the industry response to *Silent Spring*, there emerged a contest reminiscent of that between Elijah and the prophets of Baal, with each side accusing the other of defrauding the public with false signs and wonders.

For her part, Carson accused agribusiness of feeding concerned citizens "little tranquilizing pills of half truth" and of flat-out concealing what they knew about pesticide risks to make money for their bosses.[26] After discussing cheaper and safer biological controls in the last chapter of *Silent Spring*, she charged: "These extraordinary capacities of life have been ignored by the practitioners of chemical control who have brought to their task no 'high-minded orientation,' no humility before the vast forces with which they tamper."[27] She accused the scientists who took industry money for pesticide research of "serv[ing] the Gods of profit and production":[28] how, then, could their vision of the future of the ecosystem be trusted? Speaking ellipticially of the National Academies of Science National Research Council (NAS-NRC)—which had been tasked with reviewing pesticide use in the United States—Carson demanded: "We see scientific societies acknowledging as 'sustaining associates' a dozen or more giants of a related industry. When the scientific organization speaks, whose voice do we hear—that of science? Or of the sustaining industry? It might be a less serious situation if this voice were always clearly identified, but the public assumes it is hearing the voice of science."[29] She was figuring Establishment scientists as false prophets interested more in gain than the truth.

For their part, her detractors readily picked up on Carson's reverence for the holism of nature, accusing her of being a "priestess of nature,"[30] a medium for "worshippers of 'natural foods,'"[31] and of sporting a "mystical attachment to the 'balance of nature' myth."[32] They figured her as the prophet of a primitive and outmoded religion opposed to the enlightened progress of industrial science. Chemists turned polemicists such as Thomas H. Jukes from Cyanamid matched Carson's jeremiads in *Silent Spring* with parodies that exaggerated the consequences of forswearing environmental chemicals. In "A Town in Harmony," published in *Chemical Week*, Jukes conjured a stream laden with bacteria "flow[ing] through meadows where grazed cows beneath whose hairy flanks swarmed trillions of tuberculosis organisms, waiting for their milky ride that would take them to the lungs of the townspeople."[33]

Many of the industry attacks attributed Carson's marginal beliefs to her status as a woman. The gendering of the criticism of *Silent Spring* is striking, with critics attempting to negate Carson's authority by calling her a "spinster" who was "emotional," "hysterical," and "sentimental."[34] My personal favorite among these came in a letter to the *New Yorker* from "a gentleman from California" after the serialization of *Silent Spring* began. It was one of Carson's favorites, too; she read it to laughter in her National Parks Association speech in October 1962:

> Miss Rachel Carson's reference to the selfishness of insecticide manufacturers probably reflects her Communist sympathies, like a lot of our writers these days. We can live without birds and animals, but, as the current market slump shows, we cannot live without business. As for insects, isn't it just like a woman to be scared to death of a few little bugs! As long as we have the H-bomb everything will be O.K. PS. She's probably a peace-nut too.[35]

Chuckle if you like (I know I did), but this commenter was sharp enough to recognize that Eisenhower's "scientific-technological elite" were increasingly America's Establishment, outside whose walls Carson cried out as a kairotic female prophet.

Carson's prophetic opponents pursued this line of argument by arguing she wasn't really a scientist—even though she had an MS and a decade of experience in a federal biological agency. Thomas Jukes, who seems to have taken on the project of discrediting Carson as a sort of personal crusade, wrote a series of acid letters to the PSAC as it was deliberating over its pesticide report. In one of these letters, Jukes reminded the committee of a poster in the office of the FDA Division of Pharmacology that joked that you could learn the science in "three easy lessons of five years each"; Jukes then sniped, "I don't believe that Miss Carson has taken these lessons."[36] As part of the quarter-million-dollar campaign that Jukes's company helped launch against *Silent Spring*, Dr. Cynthia Westcott, a botanist and gardening writer, was paid to give lectures discrediting the science in the book. Dr. Westcott invited Carson to debate in front of the media, but Carson refused to engage in such a direct prophetic competition.[37] Carson was quite ill by that time, and she and her publisher were determined to keep that fact secret. They did not want her to appear "interested" in her own right—that is, motivated to speak not by concern for others but by selfish bitterness over an affliction that could be attributed to pesticide exposure.[38] Her friends' zealousness in defending her image as an independent public servant reminds us that her ethos was one of the main stakes in the *Silent Spring* debates.

In the CBS program, the prophetic contest was exaggerated by Dr. Robert White-Stevens's selection to represent the agribusiness position. Lab-coated, bespectacled, and mustachioed, he sat behind a desk full of bubbling test tubes and inveighed in a quasi-British accent against the "groundless" claims of "Miss Carson." As he talked, the camera cut to images of belching smoke-stacks, swarms of insects, and starving Africans. While these images were supposed to support White-Stevens's arguments about the dystopia that would result if pesticides were banned—famine and plague, a "return to the dark ages"[39]—it was hard not to connect them with his Faustian persona. On the other hand, when the camera focused on Rachel Carson, it found a pretty middle-aged woman sitting calmly in her tastefully appointed living room, elegizing the beauties of nature in gentle cadence; or, it followed her as she wandered serenely through the Maine wilderness with those ever-present binoculars around her neck. The CBS crew intercut her readings of passages from her book with beautiful natural scenes.

This ethical contrast was not the subtle business of rhetorical analysis; it was readily apparent to the viewers of the program, who in general responded negatively to the "fiendish" stereotype of the mad scientist they inferred from Dr. White-Stevens's ethical performance.[40] In this battle between the cultic priest and the kairotic prophetess, it was quite clear who the CBS producers thought merited the public trust.

Apocalypse: Carson's Construction of Uncertainty and Dystopia

It is striking how many times in the CBS program the interviewer goads a government scientist into saying "we don't know," "it's uncertain," "we can't say," or "we just don't have that information." As Kenny Walker and I have argued elsewhere, Carson stressed the uncertainty of scientists—in both the sense of their *ignorance* of pesticides' effects, and the *risks* of applying them—to negate her opponents' arguments about the effects of pesticides. Once the cultic prophets' authority was voided at the pivotal stasis of cause/effect—once it was clear that the "oracles were dumb," to steal a phrase from Bacon[41]—Carson then invited readers to step into the rhetorical gap she had created and to evaluate the risks of pesticides for themselves:

> We urgently need an end to these false assurances, to the sugar coating of unpalatable facts. It is the public that is being asked to assume the risks that the insect controllers calculate. The public must decide whether it wishes to continue on the present road, and it can do so only

> when in full possession of the facts. In the words of Jean Rostand, "The obligation to endure gives us the right to know."[42]

But Carson did not leave readers unguided at the stasis of value. She supplied graphic descriptions of cases of pesticide poisoning: squirrels dying clawing at the ground in agony, babies reduced to human vegetables by exposure to endrin, male pheasants turning into females from DDT-related endocrine disruption, rivers choked with rotting fish. The "Fable for Tomorrow" that begins the book is full of such apocalyptic imagery, and by refusing to offer the reader scientific reassurance that these nightmares could not become realities, Carson instilled fear that she hoped would lead to political action.[43]

Carson's kairotic stance afforded her more leeway to make these value judgments and calls to action than she would have had trying to make the same arguments from inside the USFWS. She was writing as an independent researcher, which freed her from the powerfully normative censorship of the scientific Establishment. And she published in a nondisciplinary genre (the trade book) that provided ample space for her vivid examples, allegories, and moral conclusions.[44]

But while she was free to make prophetic claims in her book, she could not entirely escape the strictures of the is/ought model. Her critics, as discussed earlier, used her imagery, emotional language, value judgments, and calls to action as evidence that she was not, could not be, a true scientist. The criticism stung Carson and her editors; in press materials they took every opportunity they could to remind readers of her master's degree and her years of service with the USFWS.[45] But such criticisms were inevitable given the free use Carson made of the prophetic formulae of judgment. In the last chapter of *Silent Spring*, "The Open Road," she warned,

> We stand now where two roads diverge. But unlike the roads in Robert Frost's familiar poem, they are not equally fair. The road we have long been traveling is deceptively easy, a smooth superhighway on which we progress with great speed, but at its end lies disaster. The other fork of the road—the one 'less traveled by'—offers our last, our only chance of reach a destination that assures the preservation of our earth.[46]

As elegant as these expressions of uncertainty and apocalypse may have been, I do not want to overemphasize their rhetorical impact. They represent only two threads in the complex tapestry of *Silent Spring*. It is clear, however, that CBS took these threads as its narrative through-lines. First, producers goaded Carson's opponents over and over again into admitting they did not know the ultimate effects of pesticides. Second, CBS brought the apocalyptic verbal images in Carson's book to visual life. The program is full of footage of planes engulfing fields in poisonous fog, children skipping along merrily behind a tanker truck that is spraying pesticides all over their suburban lawns, chemical plants belching black smoke. Then, there are the transgender pheasants eying the viewer with glassy stares, the choked rivers, the prone fish and birds.

The letter-writing campaign that followed the CBS program was immediate and monumental; the USDA and FDA were flabbergasted at the volume of mail they received calling for more federal oversight of pesticide use and abuse.[47] When the PSAC finally issued its report, strongly supportive of Carson and *Silent Spring*, CBS ran a follow-up called *The Verdict on Rachel Carson's* Silent Spring, which recut images from the first program with Severeid's commentary on Carson's victory. This reaction supported her ethos as a kairotic prophet calling down public judgment on the industrial scientific Establishment, and her vision of the future had won the day in the prophetic battle. It was in this spirit that she was received by the Ribicoff commission.

The Federal Testimony: The PSAC

Carson testified in front of the PSAC in January 1963, before the CBS program aired but well after a judicious public relations campaign by Houghton Mifflin had placed preprints of the book in the hands of high-ranking federal officials.[48] As early as August 1962, when Kennedy was asked about *Silent Spring*, he replied that he had already tasked the PSAC with reviewing federal spraying programs. A federal Pesticide Review Board had been sitting before *Silent Spring*, but internal documents reveal that the PSAC determined to mount a wider review, not just of federal spraying programs but of pesticide risks in general, to answer the "enlarging and justifiable national concern" Carson's articles had sparked.[49]

Kennedy's announcement touched off the most intense of the anti-Carson campaigns by agribusiness, but lobbyists had already petitioned the USDA and FDA to refute the book. A few initial statements in that direction by government officials, however, were quickly quelled byJerome Wiesner, Special Science Adviser to Kennedy, who instructed the agency heads to make only neutral statements about the benefits of public debate and the need for more research on pesticides until the PSAC could prepare its report.[50] The PSAC's internal records for the fall of 1962 contain many letters from agribusiness expressing concern about Carson's book and about what the panel's report would say.[51]

Unpublished drafts of the PSAC's report before and after it met with Carson document its sympathetic reception of Carson as a scientist-prophet. The pre-Carson drafts dealt with historical and technical claims about pesticides, which fell almost exclusively on the "is" side of the is/ought divide (the members of the PSAC were themselves scientists). The drafts after Carson's testimony, however, include much stronger statements about the effects of pesticides, statements that imply prophetic value judgments. This shift up-stasis is visible almost in the first sentences of the report. The before draft commences with a definitional claim: "Pesticides may be defined for our purposes as chemicals used for the control of insects, mites, plant diseases, weeds, nematodes and rodents." The authors go on to admit "Because these chemicals are designed to kill or upset metabolically some living organism, they are necessarily more or less toxic; it is this characteristic of toxicity and the associated inadvertent hazards to other forms of life, which has caused concern."[52] However, this admission is followed by three pages detailing the benefits of pesticides with no further mentions of their harmful effects. The after draft begins quite differently:

> Since the beginning of recorded history, man's primary concern has been the struggle for survival and the improvement of his lot. As his numbers increased, so did his powers for collective domination of his environment. Essentially, all of these advances have brought with them a degree of risk which society has accepted as an inevitable part of the price of progress.[53]

Gone are the three pages of benefits of pesticides; they are replaced instead with a caveat: "The benefits of these substances are still most apparent, but we are beginning to question some of the less obvious effects and potential risks."[54] The introduction concludes, "The Panel was arrested by the need to understand more completely the properties of these chemicals, and to determine their long-term impact on biological systems including man,"[55] These topoi—incomplete understanding, long-term "impacts," biological "systems" that include humans—are totally absent in the before draft and are worded nearly identically to the articulations of them in *Silent Spring*.[56]

Two further edits suggest Carson's prophetic influence: on page 25 the after draft discusses biological controls of pests, a recommendation that had not appeared in the original *New Yorker* serialization but that Carson had appended to the book version and that was emphasized in the CBS program. The after draft also discusses on page 28 the formation of an agency to oversee the use of environmental chemicals, a key Carson recommendation that would eventually be realized as the Environmental Protection Agency (EPA).

But most tellingly, a reference to Carson's prophetic ethos appears at the end of this and all drafts following Carson's testimony:

> Writings in the public press as well as experiences of the Panel members indicate that until Miss Carson's book there was general unawareness of the information on pesticide toxicity already available. The Government should develop the means to present this information to the public in a way that will make it mindful of the dangers without destroying its confidence in the great value of pesticides when properly used.[57]

Here, the PSAC depicts Carson as enlightening the public when government has failed it. Interestingly, this reference drew criticism from the USDA when they reviewed the draft in May, but the PSAC kept it in; as a matter of fact, the only change it made to this passage was to remove the "great" in front of "value of pesticides."[58]

Whether all these edits were solely due to Carson's influence is beside the point;. while I am interested in the reception of her specific arguments, I am much more concerned with the wider constitution of *Silent Spring* as a jeremiad against pesticide misuse. What seems certain is that as it participated in the kairos surrounding *Silent Spring*, the PSAC felt compelled to address questions of moral and political certainty regarding human interventions in nature; so doing, the committee participated in the constitution of Carson's prophetic ethos.

The Federal Testimony: The Ribicoff Commission

Carson's testimony before the Ribicoff committee was a matter of public record, so here we can see the legislators responding to her ethos in real time (Figure 7). Carson was seated in the midst of a halo of microphones that mirrored Oppenheimer's "chalkboard halo" in so many photos: the mirror images dramatize once again the contrast between Carson's role as people's prophet and Oppenheimer's role as Establishment prophet.

Senator Ribicoff opened the session by greeting her, "You are the lady who started all this."[59] This was a rough quote of Abraham Lincoln's legendary greeting to Harriet Beecher Stowe, an allusion Eric Sevareid had also chosen to open the CBS program.[60] These Carson/Stowe references, repeated over and over again in the wider constitution of Carson's ethos, confirmed her reception as a prophet of political reform.

FIGURE 7 Rachel Carson testifying before the Ribicoff committee in June 1963. The partial halo of microphones around her head mirrors Oppenheimer's chalkboard halo and positions her as a prophet of the people rather than of the scientific Establishment.

Source: Courtesy of the Associated Press.

Instead of giving the appearance of an interrogation, Ribicoff deferred to Carson again by finishing his exordium, "Please proceed as you see fit."[61] Her testimony opens with the same prophetic formulae that run through *Silent Spring*:

> The contamination of the environment with harmful substances is one of the major problems of modern life. The world of air and water and soil supports not only the hundreds of thousands of species of animals and plants, it supports man himself. *In the past we have often chosen to ignore this fact. Now we are receiving sharp reminders that our heedless and destructive acts enter into the vast cycles of the earth and in time return to bring hazard to ourselves.*
>
> This problem you have chosen to explore is one that must be solved in our time. I feel strongly that a beginning must be made on it now—in this session of Congress.[62] (my emphasis)

Carson went on to present the results of very new scientific studies on insect resistance to pesticides as well as new arguments about biological controls, confirming her scientific authority. She made a series of policy recommendations, during which she again channeled the voices of silenced Others into the inner sanctum of government: "The plight of the person affected by these poisons is pitiful. Many case histories have come to me in letters. As a rule these people can find no physician who understands their problem."[63] Later, continuing on the thread of the confirmation of her kairotic ethos, Senator Ernest Gruening asked her

> In this connection, the question that Senator Ribicoff has just asked, you probably saw a two-page spread in the recent issue of the *New Yorker* in which a lady comes into a shop with a lot of bug killers on the shelf and she says, "Don't sell me anything Rachel Carson wouldn't buy."
>
> RIBICOFF: "Do you want to comment on that?"
> CARSON: "No. I think I will let it speak for itself."[64]

The commissioners then invited her to make judgments and recommendations at the prophetic stases of value and action. Senator Gruening asked, "What would you think of creating a department of ecology that would have an overall supervision of these functions, or at least an agency of ecology in one of those departments that would try to coordinate these conflicting interests?" Carson replied, "Well, it certainly is a good objective. Whether it is feasible to do this I don't really know."[65]

Ribicoff later characterized Carson as a "true believer" and opined that no one in the chamber that day could doubt her integrity. He testified to her impact on the committee's verdict, "I have always known philosophically that one who believes is a majority because most people don't believe in anything, and here was a person who deeply believed in what she was saying."[66] The Ribicoff committee went on to recommend, as had the PSAC, the same essential policy changes that Carson had espoused—conservative use of pesticides and the formation of a federal agency dedicated to oversight of agricultural chemicals. Within a decade, part of Carson's vision for "the road less traveled" had become reality with the founding of the EPA and the banning of DDT.

We can see from both the public and the governmental reception of Carson's rhetoric that she was confirmed as a kairotic prophet dedicated to voicing the concerns of silenced citizens over and against the false prophets of the scientific Establishment. Her marginal status with respect to government science may have been a liability in her dealings with agribusiness, but it gave her the freedom to speak her conscience, and it distanced her from the accusations of disloyalty that plagued Oppenheimer's ethos as "high priest" of Establishment science. Carson also benefited from the publication of her strongly prophetic argument just as a progressive administration was coming to power.

Conclusion: The Fractured Ethos of the Science Adviser

After studying the differences and similarities in Oppenheimer's and Carson's performances of prophetic ethos, we are in a better position to understand both their honors in 1963, and the ethical trials that preceded them. In the twentieth century, the progressive and is/ought models of scientific-prophetic ethos were both operational. In times of crisis, such as the Cold War and the strontium-90 scare, scientists were called upon to ascertain the future. But when they made these predictions, the articulation of negative effects and risks unavoidably invoked evaluation; these value judgments immediately opened scientists' disciplinary ethos to charges of bias and arrogance. If the science adviser dared to make outright recommendations for action, ethical censure generally intensified, resulting in a radically unstable ethos for the scientist-prophet.

We have also seen that kairotic and cultic nuances matter a great deal in the prophetic ethos of policy scientists. Kairotic prophets such as Carson could not command the resources or authority of cultic prophets such as Oppenheimer, but by the same token they were better insulated from charges of disloyalty to the new Establishment. These ethical positions were analog, not digital; many positions were and are possible on the spectrum from kairotic to cultic. But none of them buys the scientist ethical stability. Oppenheimer fused cultic and kairotic ethē in his atomic-policy advice and statements to popular media; but this ethical configuration proved radically unstable and melted down in the face of a political regime opposed to his political stance.

A major factor in the constitution of scientific-prophetic ethos that was revealed but not pursued in this chapter was the influence of media. Carson, and Oppenheimer in his kairotic phase, used mass media to bypass official Establishment channels and engage lay polities directly in dialogue. The next chapter explores this dynamic in the media campaigns of 1980s science popularizers Carl Sagan, Stephen Jay Gould, and Stephen Hawking.

8 MEDIA, METAPHOR, AND THE "ORACLES OF SCIENCE"

At the end of the first episode of *Cosmos: A Personal Journey*, Carl Sagan stands in his trademark turtleneck before a three-dimensional glowing grid. It is a "cosmic calendar": in the upper left corner of the calendar, on January 1, a fuzzy red ball of light indicates the Big Bang; way down in the lower right corner, "in the last ten seconds of the cosmic year," a tiny speck of light sums up human history. Sagan reaches into "December" with his left hand, and the camera zooms in to frame it against a backdrop of prehistorical savannah. Two early humans crouch under a tree looking terrified at the giant finger pointing down at them from the sky. Sagan enters the frame at their scale and glances up at the giant hand, which politely withdraws, leaving him to walk the audience past those cowering primates and through our evolution. At the end of it he crouches beside the tiny, glowing square of those last ten seconds of cosmic history, encompassing "everyone we've ever heard of." He lays an early bone tool reverentially at the corner of this square, as though it were a grave, and warns the audience: "We have a choice. We can enhance life and come to know the universe that made us, or we can squander our 15-billion-year heritage in meaningless self-destruction. What happens in the first second of the next cosmic year depends upon what we do."[1]

Those who watched *Cosmos*, and over 500 million people did, know that this was not the weirdest vignette in the series, not by a long shot. Sagan emoted and gesticulated his way to becoming the most recognized scientist in the world during the early 1980s, with sold-out lectures, a regular berth on Johnny Carson's *Tonight* show, and a run of best-selling books. He drove a bright-orange Porsche with a vanity plate that read PHOBOS, the name of a Martian moon beyond which the *Voyager* spacecrafts—a project on which Sagan had consulted—had hurtled two years before *Cosmos* aired. Sagan was a bona fide scientific celebrity.

There had of course been others before him, such as Albert Einstein and J. Robert Oppenheimer, but the explosion in media coverage of science during the space race created an unprecedented niche for prophetic figures such as Sagan, Stephen Jay Gould, and Stephen Hawking, who all became just as famous—if not more so—for engaging nonscientists in dialogue about natural wonders as for doing science. Variously called "the third culture," "visible scientists," "public scientists," and most strikingly, the "Oracles of Science"—these figures have received some scholarly attention, though less than they deserve.[2] The bulk of this work has tended to chalk up the phenomenon of the Oracles primarily to personal charisma. As Giberson and Artigas argue, "They have impeccable scientific pedigrees, but it is their unusual gift for communication that has given them a platform for speaking to millions outside the scientific community, rather than the tiny audiences of specialists to whom their colleagues speak."[3]

We know by now that a great deal more than personal charisma goes into the making of a scientist-prophet. Previous work on the Oracles has acknowledged a few of these structural factors—the media, contemporary politics, or public demand—but has tended to mention them in passing or to lose sight of them during meticulous readings of the Oracles' media productions.[4] Of course, prophetic ethos is only one factor in the phenomenon of Hawking, Gould, and Sagan's ascendance and continuing popularity. But it is a major one, as intimated by Giberson and Artigas's epithet and by numerous related references in contemporary reception. Prophetic metaphors were understandably tough for reviewers to avoid when confronted with titles like *The Demon-Haunted World: Science as a Candle in the Dark*. A reviewer for another of Sagan's books, the companion to the *Cosmos* series, took exception to its "self-conscious religiosity" but nevertheless enthused "I look forward to what the mature prophet will one day have to say, for I too share his faith."[5] Likewise, Daniel Dennett, an evolutionary psychologist and one of Stephen Jay Gould's sharpest critics, called attention to the prophetic strain in his rival's essays: "Gould has certainly seen his task as one with cosmic implications, something that is especially clear in the epiphanies about the Burgess Shale in *Wonderful Life*. That makes his world-view a question of religion in one important sense, whether or not it has among its direct ancestors the official creed of his religious heritage—or any other organized religion."[6] A reviewer of Hawking's most famous book, *Brief History of Time*, identified him as a member of a "new clerisy."[7]

Following up on these hints, I examine in this chapter how Sagan, Hawking, Gould, and the polities they addressed together constituted their powerful prophetic ethos—with special attention to structural media effects related to controversy, the use of metaphor, and the uncontrollability of dialogue. I finish by asking where the women were in these media productions. But first, I provide a brief summary of the Oracles' confirming signs.

Confirmation of the Oracles

Although biography should not be the terminus of an investigation into the Oracles's ethē, it is an unavoidable way station. All three men have emphasized their callings to science and to science popularization. Gould located his epiphany as occurring at the age of five as he stood in front of the tyrannosaur skeleton in New York City's Museum of Natural History. Sagan's occurred at the same age as he gazed up at the stars shining hazily between tenement rooftops in his Brooklyn neighborhood.[8] Sagan elaborated two related epiphanies that occurred in the same year: one came after Sagan claimed to see the World War II campaigns his mother was describing to him, when his mother clarified the difference between imagination and reality for him; the other came when his father wrote out the numbers up to 1,000 for him to see and Sagan was permanently struck with "the magnitude of large numbers." About these formative instances Sagan commented that although his parents had no interest in science, "in introducing me simultaneously to skepticism and to wonder, they taught me the two uneasy cohabiting modes of thought that are central to the scientific method."[9]

Stephen Hawking's calling to science communication came when he was already well on his way to pursuing a physics degree—when he was diagnosed at the age of 21 with amyotrophic lateral sclerosis (ALS), more commonly known as Lou Gehrig's disease. Until that time, Hawking claimed, he had been an indifferent student with no clear calling in life. After his diagnosis, he fell into a deep depression. He had a series of vivid and troubling dreams, some in which he was about to be executed and some in which he was called upon to sacrifice himself to save someone else. "I suddenly realized that there were a lot of worthwhile things I could do, if I were reprieved," Hawking reported to his biographer, Kitty Ferguson. "I started working hard for the first time in my life."[10]

Humility, compulsion, and asceticism, as indexed by Hawking's comments about his physical challenges, featured in autobiographical comments by all three authors. Gould routinely made the humble claim that he was bad at math and an indifferent field paleontologist; he credited what success he had to "hard work" and an "odd way with words and a passion for ideas and facts." In one of these confessions, Gould alluded to Milton's ascetic lament "On His Blindness": "What can one do but make the most of a few lucky gifts? (They also serve who only stand and wait.)"[11]

I could not find any reviews of Sagan's media productions that touted his humility: in fact, several accused him of quite the opposite. Indeed, he frequently proclaimed science as the only true prophetic authority. Sagan boasted in *The Demon-Haunted World*, "Not every branch of science can foretell the future—paleontology can't—but many can and with stunning accuracy. If you want to know when the next eclipse of the Sun will be, you might try magicians or mystics, but you'll do much better with scientists."[12] Hawking had made a similar argument a few years before in "The Future of the Universe," where he employed the Delphic oracle as a comedic foil for the reliable predictions made by cosmologists like himself.[13]

However, what Sagan may have lacked in humility he made up for in performances of compulsion and selfless battles with false prophets. He presented his media campaign as a battle for the American mind against persistent enemies—pseudoscience, the military-industrial complex, and fellow scientists who felt "popularizing" cheapened both the scientist and the science.[14] When communications scholar Rae Goodell asked him about these trials, Sagan revealed his compulsion to press on:

> Except for what I'm about to say... I would think the negatives outweigh the positives by a large factor. But I get kids writing to me and saying; "I never realized what an exciting thing it is to be interested in astronomy; I'm very bright and I get all these good grades, and I never knew about astronomy. Now I'm going to be an astronomer." Now suppose one really good person becomes an astronomer because of such early stimulation. One really good person in astronomy can make a huge difference. It can be the most important thing I do, much more important than any of my research, to turn some exceptionally bright little kid on to science.[15]

Following up on this theme in *The Demon-Haunted World*, Sagan countered arguments that science popularization was inappropriate with the image of himself as a prophet sent to save people like "Buckley," a cabby who wanted to hear his opinion about UFO's, a man whom Sagan judged as "intelligent, curious" but led astray by his environment: "our cultural motifs, our educational system, our communications media had failed this man."[16]

Hawking's debilitating condition is a delicate subject; nevertheless, his ascetic heroics in the face of ALS remain key to the constitution of his public ethos: all but two of the 30 professional reviews I read for *A Brief History of Time* pointed out the oxymoron of Hawking's mental and physical abilities. Some have dared to argue that Hawking's celebrity is due in large part to his disease—one of these critics being Hawking himself, who once quipped, "No one can resist the idea of a crippled genius."[17] Others have pointed out that there are a lot of brilliant physicists doing cutting-edge cosmology who are quite good at explaining it to lay people,[18] but all are overshadowed by that striking media image of the brilliant mind trapped in the whirring cage of the electric wheelchair—sometimes silhouetted against a galaxy, sometimes against a chalkboard halo of thermodynamic equations.

So far, so much the same for most media celebrities, whose biographies all feature obstacles overcome on the path to achieving childhood dreams. But the Oracles were and are different in a crucial aspect—their privileged access to knowledge. The media circus surrounding them—the harping on Hawking's disability, the parodies of Sagan's diction and turtlenecks, the stirring of the evolution-versus-creation pot with each new Gould book published—would have pulled up stakes and left town if the Oracles couldn't confirm their command of methods and subjects beyond the public ken. Cosmology and evolution are both utterly insensible. We would have no awareness of their forces, no chance of using them to frame questions about our origins and meaning in the universe, without the aid of Sagan's *Voyagers*, Hawking's equations, and Gould's computer models. The demonstration of these special powers, and the bully pulpit the Oracles commanded as a result, depended on the mass media.

Media and Prophecy

Mass media—by which I indicate broadcast news and entertainment, both print and electronic—has a host of fascinating characteristics that could be brought to bear on the Oracles' ethē. Perhaps the most relevant is the nudge up-stasis that mass media tends to give scientific claims. Multiple scholars have observed that scientific claims tend to be cast in terms of lay values (such as novelty, wealth, or health) and to sound more certain when broadcast by mass media.[19] This tendency creates a ready-made kairos for prophetic evaluations and recommendations.

At least three additional features of mass media synergize with the prophetic ethos of science advisors: these are the tendency to frame science in terms of controversy; heavy trade in visual and verbal metaphor; and, the uncontrollability of mass media messages. I will explore each of these features with reference to specific examples from the Oracles' media productions during the 1980s and 1990s.

Prophetic Battles in the Media

As we discussed with respect to the media coverage of *Silent Spring* in chapter 7, the mass media tends to make events out of issues. This truism especially applies to science coverage; if there are no earth-shattering discoveries or natural disasters in a given week, the next best event is a scientific controversy. Such a media effect creates an ideal kairos for prophetic battle, in which two (or more) prophets vie for political ascendancy by trading power plays and accusations of falsehood. We saw in chapter 2 prophetic battles between Elijah and the prophets of Baal on Mount Carmel; we saw them again in chapter 4 as the Royal Society vied with clergy and courtiers for the right to determine natural knowledge pro bono publico; we saw them in the twentieth century as Carson and Oppenheimer fought other government scientists over environmental and nuclear policy. As I mentioned above, Sagan often presented himself as in a holy war with dark forces seeking to mislead Americans, particularly children, into superstition. But no scientific-prophetic battle was more salient in the United States during the last decades of the twentieth century than the "Darwin Wars," as journalist Andrew Brown termed the evolution education debates in which Stephen Jay Gould was a leading figure. These got so hyperbolic that Brown concluded "that scientific disputes are taking on much of the venom, as well as the importance, that theological disputes had in the days when theology was believed by educated people to deal with important features of the world."[20]

They started innocently enough. Gould published a paper on "punctuated equilibrium" with Niles Eldredge in 1972. Their thesis was that traditional gradualist Darwinism could not explain gaps and mass extinctions in the fossil record; a theory of long periods of evolutionary stasis "punctuated" by rapid bursts of phylogenetic change was a better explanation. Their contention would probably have flown under the media's radar as a cloakroom squabble among experts but for two things: one, Gould wrote the initial paper, and substantial portions of the ones that followed, in a colorful and accessible style uncharacteristic for the field;[21] two, he suggested that his British opponents, such as Richard Dawkins and John Maynard Smith, were misreading the fossil record not for scientific but for political reasons. He presented the British adaptationists as blindered by a Victorian capitalism that abhorred revolution the way Aristotle's nature abhorred a vacuum. Punctuated equilibrium, by contrast, was literally revolutionary. "It may not be irrelevant," Gould coyly pointed out, "that one of us learned his Marxism, literally at his daddy's knee." (He meant himself; Eldredge was a conservative.)[22]

Now, the media had a camera-ready prophetic battle. The punctuated equilibrium debate proved to be only the opening shot in a 20-year campaign fought on two fronts: in the public eye, it was a war to settle who had the right to opine on humanity's origins—preachers or biologists; within the hallowed halls of biology, it was a war over who had the right to call themselves Darwin's true prophetic heirs—British adaptationists and sociobiologists, or the "Young Turks" Gould, Eldredge, and Richard Lewontin.

With Lewontin Gould wrote his controversial "Spandrels of San Marco" paper in 1978, which used an architectural analogy to argue that mutations sometimes yielded auxiliary structures that were adaptively neutral but might become adaptive in future generations if selective pressures changed. The argument was controversial not so much because of *what* Gould argued but because of *how* he argued it: again, he drove the argument with a metaphor borrowed from another field, and again, he charged adaptationists with blindness to their own political biases.

Gould presented this paper at the Royal Society because Lewontin hated to fly, and the reaction was immediate and vituperative. His goat thoroughly gotten, Gould demanded the conference organizer step clear of the podium so that everyone could see the Society motto embossed on it, *nullius in verba*; that was all he was doing, Gould protested, questioning the traditional authority of gradualism—wasn't that what scientists had sworn to do since the founding of the Royal Society?[23]

The response from Gould's opponents was spirited, to say the least. In critical responses to this and other Gouldian attacks on adaptationism, John Maynard Smith claimed Gould's "mystical" explanations "turned [his] stomach"[24]; Daniel Dennett accused him of cherishing supernatural "skyhooks" that shaped evolution toward its telos in humanity.[25] Jonathan Alcock attacked Gould's unorthodox style; he claimed Gould made the same six suspect moves in each of his essays criticizing sociobiology (a favorite whipping boy of Gould's, who thought evolutionary explanations for social phenomena were reductive):

Step 1. Advertize the scholarly credentials of the author.
Step 2. Characterize persons with opposing views as bumpkins.
Step 3. Demolish a strawman afflicted with "genetic determinism."
Step 4. Make a small (and insincere) concession to opposing views.

Step 5. Proclaim allegiance to all that is politically correct and morally desirable.
Step 6. Assert that behavior *X* is immune to adaptationist analysis.[26]

Alcock sardonically summed up Gould's ethical posture in these articles as follows:

> In other words, persons who study the possible evolutionary basis for human aggression and sexism are genetic determinists who would thwart attempts to change our world for the better. In contrast, Stephen Jay Gould abhors violence, he speaks out against sexism, he despises Nazis, he finds genocide horrific, he is unfailingly on the side of the angels. Who can argue with such a person?[27]

Alcock was not the only one to pick up on the evaluative slant in Gould's writings. Peter Atkins complained that Gould had an obsession with explaining nature: "A great deal of the universe does not need any explanation. Elephants, for instance. Once molecules have learnt to compete and to create other molecules in their own image, elephants, and things resembling elephants, will in due course be found roaming through the countryside."[28] But as Atkins—who almost certainly did not believe that molecules "learned" or "created" anything "in their own image"—learned the hard way, it was tough to fight Gould without getting sucked into his prophetic rhetoric. Gould rubbed in this irony with Daniel Dennett and Robert Wright in 1997, when he first charged them with "Darwinian Fundamentalism" and then dismissed their rejoinders as a case of "'When you have nothing to say, say it louder'—a tactic that got 450 prophets of Baal into terminal trouble with Elijah."[29]

Gould started using his essays for *Natural History* magazine as a bully pulpit for his side of the argument; these were collected in *Ever Since Darwin* (1980), *The Panda's Thumb* (1981), and later volumes. Not to be outpaced, Dawkins, Dennett, and many on the other side began selling their views to trade publishers with the aid of agent John Brockman, who was capitalizing on an increasing appetite for popular science among educated readers—the "top 10%," as Gould was fond of calling them.[30]

But the Darwin Wars really heated up when news headlines on the controversy, such as "Recent Studies Spark Revolution in Interpretation of Evolution," caught the attention of fundamentalist Christians ever on the lookout for a new chink in Darwin's armor.[31] They read Gould's popular accounts and eagerly latched onto the clear explanations of what had previously been esoteric evolutionary arguments. Gould thus found himself, overnight and against his will, the darling of campaigns to teach creation science and/or intelligent design in the public schools. "Harvard scientists agree: Evolution is a hoax!!!" crowed one such broadsheet.[32]

This mix of polities was bound to combust. Horrified at the ground the creationists had suddenly gained in public debate, adaptationists blamed Gould. They charged him with employing rhetorical tricks to mislead his popular readership, which was in a position to understand neither (a) how minor his objections really were in the overall scheme of Darwinian evolution, nor (b) the technical arguments the adaptationists needed to make to refute them.[33] But just as Gould's opponents had found it hard to resist adopting his argumentation techniques in countering his stance, many of them found themselves raising their voices to match his in the public sphere: thus, Dawkins's *The Blind Watchmaker*, Helena Cronin's *The Ant and the Peacock*, and Dennett's *Darwin's Dangerous Idea*. The appeal was not just the money to be made (and the money was considerable; all Gould's books were *New York Times* best sellers, and *Blind Watchmaker* sold over two million copies). This was also a chance to circumvent official channels of scientific communication and engage voters directly in dialogue on issues related to science funding and education. Among these new channels, none seemed more enticing than the visual media—film and television—for the wider play they gave to metaphorical explanations of science.

Media and Metaphor

As we have seen in previous chapters, prophets favor figurative modes of communication—metaphors, fables, parables. Francis Bacon's theory was that these modes captured both imagination and reason thus having a better chance of moving the will than logic alone. Media scholars agree with him that far: for instance, a recent review found that "dramatic" rather than direct presentations of public health messages such as "stop smoking" or "get more exercise" seem more effective in changing behavior.[34] Exactly why they work better is unknown, but that they do suggests an affinity between dramatic media presentations and prophetic recommendations.

In this section we consider two dramatic scientific presentations from the kairos in question: Carl Sagan's *Cosmos* and Errol Morris's filmic adaptation of Stephen Hawking's *A Brief History of Time*. Specifically, I examine the interaction of visual metaphors and prophetic divination.

Cosmos was both a television and a print media phenomenon, like *A Brief History of Time* after it. A companion *Cosmos* book was released in the fall of 1980 when the series aired. Filmed by KCET Los Angeles, the 13 episodes ran over their $8 million budget, an astronomical figure for public television in those days. Sagan and his co-creators, Ann Druyan and Steven Soter, spent most of that budget on special effects, but due to the wide range of topics covered—not just cosmology but also evolution, genetics, and nuclear science—they did also pay consultants, among them Stephen Jay Gould.

Television runs on stories and pictures—or myth and mimesis, in the critical vocabulary of media scholar Roger Silverstone.[35] Science, by contrast, tends to work with invisible causes whose effects are registered by sensitive instruments in symbolic markings so abstract that usually only trained specialists can decipher them. Further, these causal explorations often continue for years before they are ready to be transformed into a narrative account, whereas television tells stories in an hour or less.[36] These different ways of framing space-time in the sciences and mass media have been called "chronotopes," and Jordynn Jack has demonstrated that scientists have to adjust to unfamiliar chronotopes when they try to explain their work to lay people.[37]

Therefore, it is not surprising that when science is adapted to the chronotope of broadcast educational television, it is simplified in several ways. One way is through teaching or "exegetical" metaphors.[38] Just as Protestant prophets in the seventeenth century used metaphors to explain the Bible, media producers use exegetical metaphors to compare natural and scientific processes to familiar processes, such as "dancing," "choosing," and "sleuthing," to name just a few exegetical metaphors our Oracles or their reviewers have employed.

In addition to metaphor, television producers routinely use other transformations to make science more palatable, according to Silverstone.[39] Anthropomorphic transformations are metaphorical; they compare scientists to heroes by showing them behaving in mythically heroic ways, such as gazing through telescopes at the sky, swirling test tubes full of colorful chemicals, or striding through uncharted jungles; these scientists are almost always solitary men, in defiance of the resolutely communal nature of actual scientific work.

Technomorphic transformations are metonymic; they reduce science to symbolic parts of it—particularly, its instruments. Technomorphic transformations occur whenever the camera gawks at delicate-looking microscopes, behemoth particle accelerators, shiny spectrographs, and the like. During these reveries, the audience is encouraged not to comprehend the mechanisms but to admire them as if they were butterfly wings, beehives, or other inexplicable natural wonders.

Semiomorphic transformations focus on the output of these machines and scientific procedures. The inscriptions, gram stains, and blips on screens that are the products of science are occasionally demystified by voice-over explanation, but more often they are presented simply as an inscrutably wonderful human code that rivals the hieroglyphics of nature—dunes or ant columns or mud cracks.[40]

All three transformations relate to prophetic Divination or Confirmation: the scientist-prophet as the lone hero with special access to messages in the "wilderness," the divinatory techniques for inscribing those messages. Both *Cosmos* and *Brief History of Time* make heavy use of these transformations. To return to the first episode of *Cosmos*, "On the Shores of the Cosmic Ocean," we first meet Sagan striding through the wilderness to a wild seafront, where he picks up a dandelion seed. It is transformed through special effects into a "starship of the imagination" that ferries Sagan and his viewers out into the universe to look for the origins of human life. Sagan appears again at the helm of this ship, an altar-like console of glowing lights. The white walls of the ship vault into gothic arches behind him, and Vangelis's hymnic soundtrack plays reverentially in the background. Where a stained-glass window or crucifix would be located in a Christian church, Sagan's sanctuary is dominated by a giant window into space, spangled in succession with galaxies, nebula, and pulsars. As Sagan explains the cosmic mechanics of these phenomena in more detail, he occasionally passes his hands through the air over the floor of the ship, and another "window" opens there, depicting what he describes in more detail.

These structures and gestures are visual metaphors that transform the investigatory work of science into dramatic acts of Divination. Instead of showing Sagan seated at a bank of computer monitors—the primary setting in which his disciplinary ethos as an astronomer is constructed on a daily basis—the directors transform the radio or optical telescope Sagan would be using into a church-like vehicle that takes both Sagan and the audience out into the cosmos. There, they wonder at structures that would be impossible to view with the naked eye at their actual scales and luminosities. The scientist is the priest of this sanctuary, exhorting his flock to wonder—a striking anthropomorphic transformation of Sagan's disciplinary ethos, and one that many of his scientific colleagues reacted negatively to.[41]

Cosmos might seem particularly susceptible to these transformations because of Sagan's personal predilections toward the emotional and mystical in media representations. After all, he claimed in *Demon-Haunted World*, "Science is not only compatible with spirituality; it is a profound source of spirituality."[42] Yet, these prophetic transformations appear to be endemic to media representations of cosmologists. Errol Morris's film adaptation of *Brief History of Time* is equally full of them.

Hawking reportedly demanded payment up front for his involvement in the documentary,[43] which devotes a surprising amount of screen time to framing his hand as it clicks the operator for his voice computer, composing sentences that the machine then plays back in its superhuman monotone: "Which came first, the chicken or the egg? Did the universe have a beginning, and will it end? What is the nature of time, and why do we remember the past but not the future?"

The anthropomorphic transformations of Hawking are stunning. At one point Morris cuts from a simulation of a star collapsing upon itself to create a black hole—another layer of metaphor as a singularity is not a hole at all but an infinitely dense point where the laws of physics as known cease to work—to an image of Hawking crumpled in his wheelchair with the stereotypical chalkboard halo. The film ends with the famous "mind of God" passage that also ends the book, read in Hawking's computer voice:

> If we do discover a complete theory, it should in time be understandable in broad principle by everyone, not just a few scientists. Then we shall all, philosophers, scientists, and just ordinary people, be able to take part in the discussion of the question of why it is that we and the universe exist. If we find the answer to that, it would be the ultimate triumph of human reason—for then we would know the mind of God.[44]

During this passage, the camera lingers on a backlit rearview of the wheelchair again with its vanity plate: "STEPHEN." Morris's transformations of Hawking verge on apotheosis.[45]

Hawking's initial divinatory questions are answered over the course of the film via computerized visualizations of space-time, many of which were not only Hawking's representations but were in fact *the very forms in which he apprehended the theories*: particles "dancing" with antiparticles at the event horizon of a black hole, "cones" of space-time development flattening into "bowls" as the theoretical need for a starting point for the universe vanishes. A teacup shatters on the floor and then, through the magic of film, reassembles itself to illustrate Hawking's question about the asymmetry of memory.

More than just media transformation is at work in these elaborate metaphors. According to Hawking, the metaphors came first and the equations later as a result of his great difficulty writing. He first developed a holistic visual symbol system for representing theoretical relationships. Then, he worked out the equations with the assistance of graduate students in a dialogue reminiscent of those between the Pythia and her *prophētōn*: he would tell his assistants what proof to attempt, they would work out the math and bring it to him, and then he would correct them or point out a test case and send them back to try it again; this process would continue until Hawking was satisfied with the mathematical representation of the metaphorical vision he had had, or until he had changed his mental model to fit the mathematical results.[46]

But Hawking is not the only Oracle, or the only scientist, to employ metaphor as a fundamental of scientific inquiry. Many of the transformations performed in *Cosmos* and *Brief History of Time*—and indeed in *Connections, NOVA,* and many of the other broadcasts from the "golden age" of public science TV—simply reenact transformations that occurred much earlier in the minds of the scientists themselves as they sought to understand invisible and unfamiliar natural processes. Through their instruments, equations, and mental models, scientists made these processes legible, they objectified them, *they created them* as far as their polities were concerned. But prior to the experimentation and instrumentation, metaphors guided the scientists' inquiry—suggested where to look and what to look for.

Ken Baake has performed an in-depth study of this process with the complexity theorists at the Santa Fe Institute. He discovered that they employ "theory constitutive" metaphors to help organize and direct their mathematical investigations: "complexity" itself is a metaphor for the operations of nonlinear equations. Other powerful theory-constitutive metaphors have included "computer" for the brain (and vice versa), "particle" and "wave" for the behavior of light, and of course "super strings," "black holes," and the other dramatic metaphors of cosmology. These metaphors do more than just help the scientists explain their ideas to others—they suggest research approaches, questions, and methods.[47] For instance, although cosmologists argue that a "black hole" is really not a hole at all, regardless, taking the metaphor of "holes" seriously led Karl Schwarzschild and later cosmologists to the theory of "wormholes," linkages within or among universes that may be "held open" by dark matter.

When these theory-constitutive metaphors are in turn employed as "exegetical" metaphors to explain science to the public, we get a sort of epistemological and rhetorical "black hole": the divinatory metaphors used to answer questions about the natural world become the very way we experience those insensible aspects of the natural world.[48] The production and reception of natural knowledge collapse into a singularity, and scientists *create* the natural world, for political purposes, through their equations and inscriptions. As Andrew Brown once mused on his attempt to understand George Price's algorithms for human altruism: "The innumerate reader must force himself past thickets of equations in search of the enchanted castle of truth—only to find that the equations were the truth."[49]

Just like Brown, we as readers and watchers of the Oracles participated in the constitution of the worlds they created for us. Before reading *Brief History of Time*, for instance, each of us lived in a certain mental model of the universe. What we understood of physics conditioned our day-to-day behavior, particularly around cliffs and heavy objects. After reading and understanding (in whatever way we did) Hawking's metaphors in *Brief History*, that mental model changed. I remember that when I read the book in college, time got very strange for a while, and I imagined space not as an inky vacuum or a star-spangled dome but as a giant black rubber sheet bowed into stretchy wells by the mass of planets. In this way, as I have argued elsewhere regarding media coverage of science, metaphorical arguments such as Sagan's and Hawking's, if they are persuasive, *literally change the world we inhabit.*[50] Or, in the words of one reviewer of both Hawking's book and James Gleick's book on chaos theory, "[B]etween Gleick's butterflies and Hawking's black holes, after reading these books you'll never look at the world the same way again."[51]

We trust the Oracles to lead us in this constitutive dialogue of the imagination because we are awed by the transformations we witness through our television sets—the exploding novae, the "black holes," the gleaming dials and test tubes, the droning of impossibly complicated electromagnets. Television, which is still treated as a transparent "window" on scientific activity,[52] is quite clearly anything but. It has become itself another piece of prophetic apparatus in the dialogue the scientist-prophets engage us in, and the ways in which it shapes this dialogue are not incidental but fundamental to prophetic ethos in the media age.

"Irenics": Prophetic Dialogue with the Oracles of Science

"Who Speaks for Earth?" the final episode of *Cosmos*, is also its most stridently prophetic. It opens with starry space and a voice-over of Sagan reading Deuteronomy 30:15, in which God makes his covenant with the Israelites through his prophet Moses: "I have set before you life and death, blessing and cursing: therefore choose life, that both thou and thy seed may live." With the word "seed," the dandelion-seed-shaped Ship of the Imagination spins into view and out into space beyond.

Early in the episode, Sagan makes clear what he believes his viewers' "choice" to be, and he does so using the classical judgment formula, just as Moses did: "As the ancient myth makers knew, we are children equally of the earth and sky. In our tenure of this planet, we have accumulated dangerous, evolutionary baggage—propensities for aggression and ritual, submission to leaders, hostility to outsiders, all of which puts our survival in some doubt."[53]

He goes on to speculate about what aliens would think of our world and our decisions. Extraterrestrial life was a perennial obsession with Sagan, leading to his involvement in the Search for Extraterrestrial Intelligence project, his writing the novel (and the story outline for the ensuing film) *Contact*, and his insistence on a hieroglyphic greeting tablet, the "golden record," accompanying *Voyager* should it encounter alien intelligence.[54] Now, in "Who Speaks for Earth?", Sagan segues from musing about the perspective of alien evaluators into a "dream" in which he observes an alien planet that suddenly goes dark. Mystified, he consults an electronic "Book of Worlds" via his ship's console and discovers that the planet destroyed itself through nuclear warfare. Overcome with dream panic, he sets course for home, where he arrives just at the moment that the peoples of Earth also annihilate themselves in nuclear war.

Sagan returns from this jeremiad to Earth's wilderness, from which he calls out for his viewers to consider the perils of nuclear proliferation before it is too late. But in the program's last moments, he shifts to isaiad while we watch hopeful images of happy schoolchildren:

> We have mapped the molecular machines at the heart of life. We have discovered a microcosm in a drop of water; we have peered into the bloodstream and down on the stormy planet to see the earth as a single organism. We have found volcanoes on other worlds and explosions on the sun, studied comets from the depths of space and traced their origins and destinies; listened to pulsars and searched for other civilizations... It has the sound of epic myth, but it is simply a description of the evolution of the cosmos as revealed by science in our time. And we, we who embody the local eyes and ears and thoughts and feelings of the cosmos, we have begun at last to wonder about our origins—star stuff contemplating the stars, organized collections of ten billion billion billion atoms, contemplating the evolution of nature, tracing that long path by which it arrived at consciousness here on the planet earth, and perhaps throughout the cosmos.[55]

Sagan is now back at the seashore where he began the series, and he again picks up a bit of dandelion fluff as he ends his narration: "Our loyalties are to the species and to the planet. We speak for earth. Our obligation to survive and flourish is owed not just to ourselves but also to that cosmos ancient and vast from which we spring." And with that, he lets the seed fly again. Our final image of him is as the captain of the Ship of the Imagination, sailing off into the cosmos alone to bring back more wisdom on our behalf.

Giberson and Artigas argue that each Oracle had a "big project," the accomplishment of which required engaging American readers/viewers in dialogue. Sagan had three interrelated projects: nuclear disarmament, funding for space exploration (including the search for extraterrestrial intelligence), and funding for public science education. He had differing levels of success engaging American polities in dialogue on these issues. He was made the spokesperson for the antinuclear Union of Concerned Scientists and issued "nuclear winter" warnings at media events and in a high-profile congressional testimony in 1983;[56] similarly, he was invited to speak on topics related to nuclear disarmament and foreign affairs both in the United States and abroad.[57] Although these invitations signal the beginnings of dialogue, subsequent downgrading of the severity of nuclear winter to "nuclear autumn" by atmospheric physicists took a great deal of urgency from the debate.[58] In an update to *Cosmos* filmed in 1991, Sagan was still urging action on disarmament, but even he admitted that public fear of nuclear holocaust had greatly diminished since *Cosmos* first aired.[59]

By contrast, his arguments for the public funding of space exploration rode the crest of a wave of renewed interest in the subject, as support for NASA's shuttle program injected fresh life into the agency after the mothballing of the Apollo missions. However, this apparent victory for Sagan's prophetic message must be tempered with the knowledge that most funding for space exploration technologies came from military programs aimed ultimately at nuclear attack and defense systems, such as Star Wars.[60] This complicated situation is a good reminder that prophets' messages must confirm underlying currents in public discourse to be heard, and that a prophet cannot unilaterally determine the outcome of a dialogue that he begins with the polity.

Finally, the dialogue about scientific literacy that Sagan started in earnest with *Cosmos* and finished with *Demon-Haunted World* has an interesting and poignant constitution history. Sagan was a frequent lecturer at schools; as we saw above, he claimed one of his primary missions was to encourage children to become scientists. Of the six or seven thousand fan letters he reported receiving after *Cosmos* aired, he proudly singled out the hundred written by parents who claimed their toddlers insisted on watching *Cosmos* even though they could not possibly understand the arguments presented.[61] One Amazon reviewer of the *Cosmos* DVD re-release recounted a public lecture in New York that the reviewer attended as a child and at which Sagan picked him out of a sea of raised adult hands to ask the evening's first question.

In keeping with Sagan's clearly received educational focus, a prominent feature of *Demon-Haunted World* was its so-called baloney-detection kit, or "skeptics toolkit." Sagan promised his readers that these were the techniques that scientists themselves used to divine "the occasional straw of truth awash in a great ocean of confusion and bamboozle"[62]; now, he was placing them in readers' hands as well so they could protect themselves from the deceptions of pseudoscience and government newspeak. These tools included a range of critical thinking precepts, such as "consider the source" and "quantify where possible," but they also contained a list of argumentative fallacies to watch out for, tricks such as "slippery slope" and the post hoc fallacy.

In spite of this Promethean gesture, some of Sagan's readers called baloney on him in turn. Critical thinking was great, but it wouldn't really bridge the gap between lay culture and the specialized work of scientists like Sagan. David Tietge pointed to a passage in *Demon-Haunted World* where Sagan claims that the main difference between shamanism and quantum mechanics is that we can "verify that quantum mechanics works." But the verification procedure he proposed was comically complicated: step number one was to acquire 15 years of specialized training in physics. Tietge concluded, "What he means is that *scientists* who are not quantum specialists can use this method to verify the conclusions that the theory predicts; the rest of us have to rely on faith in these scientists that what they tell us is true and accurate."[63] Thus, Sagan's arguments fell once again on his authority as a physicist and so emphasized rather than diminished the cultural gap between him and his readers. In this light, Richard Lewontin stressed the irony of the baloney-detection kit's "don't rely on authority" tool in his review of *Demon-Haunted World.*

> It is certainly true that within each narrowly defined scientific field there is a constant challenge to new technical claims and to old wisdom. Young scientists on the make will challenge a grey-beard and this adversarial atmosphere for the most part serves the truth. But when scientists transgress the bounds of their specialty, they have no choice but to accept the claims of authority, even though they do not know how solid the grounds of those claims may be. Who am I to believe about quantum physics if not Steven Weinberg, or about the solar system if not Sagan?[64]

In other words, Sagan's solution for breaking down the boundaries between science and lay culture was one that even other scientists could not hope to enact.

A wider view confirms this narrow impression that Sagan's isaiad for science education fell on deaf ears: federal funding for public science education has not significantly increased since the late 1980s. And yet that is not the whole story for scientific literacy and the cultural effects of programs such as

Cosmos. The series remains the most widely watched public broadcasting program of all time; the companion book was on the *New York Times* best-sellers list for over 70 weeks. *Cosmos* thus participated in a kairos in which, by some measures at least, general scientific literacy has been slowly rising despite a lack of science education funding.[65] As an interesting aside on this point, roughly a third of the 269 reviewers of the DVD rerelease on Amazon claim that *Cosmos* made a big impact on them as children; many even claim to have chosen a scientific career as a result of watching the series—anecdotal but remarkable.

Again, these prophetic dialogues Sagan engaged his audiences in do not frame questions about the "effect," "impact," or "reception" of Sagan's messages as much as they do questions about the constitution of covenant values relating to science, literacy, and origins in late twentieth-century America. Likewise for Stephen Hawking's "big project," which could be summarized as the demystification of the origins of the universe. Hawking foresaw this explanation arriving via a Theory of Everything (ToE), a grand unification of quantum mechanics and relativity that would explain the behavior of every atom in the universe and foster social unity.

Sagan expressed a related hope at the end of *Demon-Haunted World*, and as we will see shortly, Gould longed for political rapprochement among disciplines and factions. If our twentieth century Oracles could be said to share a gospel, harmonious political dialogue would be a good candidate.

But the dialogue over political rapprochement that Hawking engaged audiences in with *Brief History of Time* took the same ironic turn that Sagan's did. While many reviewers of the book praised the "clarity" and "lucidity" of Hawking's writing (epithets also systematically applied to Gould and Sagan's prophetic efforts), nearly half complained about not being able to understand it. And some of those critics perceptively pointed out that the efforts to clarify the physics via metaphor only emphasized the divide between scientists and lay readers. One reviewer claimed that books like Hawking's "preach[ed] to the converted," i.e., to other scientists;[66] another reviewer expanded on that notion, writing, "I don't think we read these books for the answers, and anyway the answers would be no help. Cosmic string, dark matter and space-time singularities are to most of us mysteries just as unencompassable as say Alpha, Omega and the Seraphim. In both cases we are content to take the word of the high priesthood."[67] Maureen Madison pointed out that Hawking undercut the prospects of social unity by the very language he chose in the "mind of God" passage—distinguishing "philosophers" and "scientists" from "ordinary people."[68]

These critical reactions extended to Sagan's and Gould's messages as well. One reviewer of *Cosmos* feared that its gorgeous cosmic visualizations tended "to create an awe of science rather than an understanding of it."[69] The reaction to Gould's *Wonderful Life* showed a similar ambivalence toward his attempts to "popularize" paleontology. One reviewer praised Gould's divinatory skill: "With skill and passion, Gould takes this mute collection of fossils and makes them speak to us."[70] But another employed the same divinatory analogy to express despair at understanding the dense paleontological explanations: "Stephen Jay Gould has the extraordinary ability to make stones talk, but what they have to say almost makes you wish that he had never tried."[71] A third reviewer directly addressed the boundaries between scientific and lay polities that the other two hinted at: "We must ask forgiveness from the scientists for putting the wrong questions to them and for misrepresenting their findings; but in return they should acknowledge how real is our esteem for their work and how genuine our disappointment at their explanations of it."[72] In this vein, Gillian Fuller found that while the body of Gould's *Natural History* essays was "projective" and "openly dialogic," nevertheless, his essays tended to lapse into didactic monologue at the end, reinforcing Gould's superiority over his readers.[73]

Ironically, then, the Oracles' stated attempts to break down political boundaries between the sciences and the lay public served in part to reinforce those boundaries. This irony can be explained via the prophetic function of Confirmation: demonstrations of special knowledge and special access to the wonders of nature reinforce the extraordinary calling of scientist-prophets. If the knowledge they brought us back from the "wilderness" were common sense, we would have no need for it or for them. Their authorization rests on the arcanity of their knowledge. This is the distinction between prophetic and expert ethos. Expertise casts the special skills or knowledge it commands as fundamentally attainable by its audience; prophetic ethos casts its special access as practically unattainable.

Just such a conundrum plagued Gould's "big project," which initially sought to break down boundaries between science and politics but shifted later to endorse such boundaries as necessary for political peace. In *Panda's Thumb* Gould laid out his manifesto: he argued that if people saw science for what it really was, a flawed human activity nonetheless pregnant with "exultation *and* explanation," they would feel empowered to make tough political choices themselves instead of leaving these decisions to traditional authorities—whether scientists, pseudo-scientists, or "preachers and politicians."[74] In a gorgeous little passage strongly reminiscent of Robert Boyle's "occasional reflections," Gould introduced his vision:

> In an old stereotype (not followed nearly so often as mythology proclaims), the natural history essay restricts itself to describing the

> peculiarities of animals—the mysterious ways of the beaver, or how the spider weaves her subtle web. There is exultation in this and who shall gainsay it? But each organism can mean so much more to us. Each instructs us; its form and behavior embodies [sic] messages if only we can learn to read them.[75]

The "message" Gould was divining for his readers in his fables about pandas and the foibles of past scientists was that Americans should examine and discard many of the traditions that shaped their present worldview—for these, too, would soon change or vanish under the selection of history.[76] This was Gould's revolution.

By the end of his career, however, Gould was calling not for revolution but for "irenics." In *Rocks of Ages* (1999), he proposed his own variation on Hume's Guillotine that he called NOMA, "Non-Overlapping Magisteria." NOMA meant that lay polities would respect scientists' right to conduct their inquiries as they saw fit, and scientists in turn would respect the rights of organized religion and popular systems of morality to determine the goals and ends of their inventions. What caused this about-face?

Of the three Oracles considered in this chapter, Gould's rhetoric over the course of his career has received the most scholarly attention. An entire volume was dedicated to rhetorical analyses of the "Spandrels of San Marco." Critics—both admirers and detractors—of these early writings received loud and clear both Gould's message about the revolution and his prophetic stance in uttering it. We saw that pattern clearly in the responses of peers who chose to engage Gould in prophetic battle, above. But rhetoricians noted it as well. Jeanne Fahnestock located the central argument of "Spandrels" at the stasis of value. Moving to this stasis often indicates a shift to prophetic ethos, and in fact, Davida Charney found that Gould's evaluative arguments provoked the most opprobrium from biologists reading the article; they criticized Gould for being arrogant, unfair, "unscientific," and promulgating "bullshit" in making those particular claims—all the typical charges against disciplinary ethos that follow when scientists cross the is/ought boundary.[77]

Here, I need to take a brief digression to explain how value judgments and polities interact to create differing effects given the is/ought model. Arguments about values or actions generally signal prophetic ethos when the scientist is speaking to nonscientific polities. However, when a scientist addresses his own polity, as Gould did in "Spandrels," whether or not he is received prophetically depends on the particular values he applies to the evaluation. If the scientist evaluates using technical scales of size, mass, force, and so on—by arguing *X* is weaker or smaller than *Y*, for instance—these technical value judgments are generally received as "is" claims—facts and definitions.[78] In addition, if the scientist applies field-internal values such as "accuracy," "intrinsic interest," or "importance to the field,"[79] he most likely will not be perceived as prophesying; many standard scientific opinion pieces and state-of-the-field addresses make such value judgments, as Gould reminded the rhetoricians who contributed to the "Spandrels" analysis volume.[80]

However, if the scientist judges the actions of his peers using *field-external* values sourced from larger or rival polities—values such as human rights issues, planks in the feminist platform, or tenets of environmental advocacy—then he is likely to be treated by his peers as performing prophetic ethos, thereby increasing his risk of censure. As we saw in Charney's study, Gould's moves to judge his field as conservative and biased earned him criticism from his peers. Daniel Dennett certainly and roundly censured Gould for the "numerous" political agendas he identified in "Spandrels" and other essays.[81] Dennett called the *Natural History* essays in which Gould first took these agendas public his "bully pulpit"[82] and denounced "Spandrels" for giving ammunition to the "Darwin-doubters."[83] And indeed, there is evidence in that essay to support Dennett's reading that Gould was inviting readers outside paleobiology to evaluate the state of the field along with him: Gould opened two sections with simplistic definitions of evolutionary terms that every one of his peers would already know; further, the opening of the paper was a long digression into architecture; in fact, one of Charney's expert readers referred to these rhetorical moves as "pedagogic,"[84] which provides an important hint that Gould was aiming "Spandrels" in part at an external, less-expert polity.[85] John Maynard Smith claimed he would not even have stooped to address Gould's arguments "were it not that he is giving non-biologists a largely false picture of the state of evolutionary theory."[86] It seems clear in Gould's case that his peers constituted his externalist evaluations of his field as inappropriate prophetic utterances.

As indicated by Maynard Smith's comment, the prophetic tone of "Spandrels" was hardly anomalous for Gould; he frequently crossed disciplinary boundaries to adopt prophetic ethos. Gillian Fuller argued that his *Natural History* essays "narrativized" the history of his field, turned it into a drama with flawed heroes and rising and falling action. Fuller argued Gould used this rhetorical strategy to "recast...theoretical issues in science into moral political issues."[87]

Giberson and Artigas observed that Gould built up to these value judgments only after painstaking argument at the lower stases. Of the ending of *Wonderful Life*, in which Gould laid bare the moral of his tale—that human life is utterly contingent and therefore should be lived gloriously free of the constraints of tradition and creed—they wrote, "These deeply philosophical conclusions—part science, part poetry, part personal religion—flow so naturally from the eloquent exposition that led up to them that it is easy to think they are also a part of the science under discussion. And there lies their strength."[88] This is the old progressive model of scientific-prophetic ethos, alive and kicking in Gould's essays despite the harsh criticism his peers imposed upon him for violating disciplinary norms. But Charles Bazerman has defended Gould's rhetorical strategy as more canonical than his critics have been willing to admit. After all, science is natural philosophy, and "philosophy is at root a wisdom literature."[89] Seen from this perspective, Gould's early rhetoric is a not an anomaly but an extension of a progressive prophetic tradition stretching back through Boyle's and Bacon's writings to find its roots in the ancient wisdom literature of Solomon and Heraclitus. Gould is in these works a prophet calling for revolution in the politics of his field and in public awareness of them.

How, then, to explain *Rocks of Ages*, in which Gould resurrects toward the end of his life a form of Hume's Guillotine? Surely, David Prindle argued, Gould must have realized his would be the first head to roll based on his history of prophetic rhetoric. Prindle read in this turn a man whose embattled experience in the public sphere had impressed upon him the value of "cultivating social peace."[90] I believe this brief conclusion merits further explication: specifically, I believe Gould's seemingly abrupt shift from advocating the progressive model to advocating a version of the is/ought model can be substantially explained by the way in which he lost control of his debate with creationists.

I have already outlined the process by which Gould's arguments on punctuated equilibrium were first picked up by the mass media and then reinterpreted by creationists and the Intelligent Design movement. Gould's response was to give expert testimony for the prosecution in *McLean et al. v. Arkansas*, the 1981 lawsuit brought in Arkansas State Court against the "Balanced Treatment" Act 590, which mandated equal time for creationist and evolutionary explanations of life origins in the public school science curriculum. The prosecution's strategy, of which Gould was the linchpin, was to show that "balanced treatment" in fact pitted a scientific curriculum against a religious one. Gould listed several contradictory predictions that creation science and evolutionary biology each made about the origins and development of life, thereby framing a zero-sum game. Not only did Gould argue convincingly that evolution was the winner in each case, but he also pointed out that every time creation science met with a developmental issue it could not resolve, it appealed to God as the solution: "I think they proceed by misquotation, by selective quotation, and by invoking supernatural intervention to produce the basic kinds of life, all of which are not only unscientific, but represent skill and rhetoric rather than science."[91] In conclusion, Gould contended, creation science was religion, not science, and as such had no place in a secular curriculum. Arkansas lost the suit 7 to 2 and had to void Act 590.

But the creation v. evolution controversy was just getting fired up. Other school boards followed suit, including Dover, Pennsylvania, in 2004. Gould's testimony in *McLean v. Arkansas* turned out not to be a coup de grace but rather the first volley in an long and exasperating dialogue with creationists: he published articles in *Science for the People* and, naturally, in *Natural History*; he lectured on the creation/evolution debate across the country; and, he dedicated a section of *Hen's Teeth and Horses' Toes* to expressing his "sadness" that the debate would not expand coverage of cosmology and origins in public schools but would rather constrain the teaching of evolution.[92]

Once a prophet has lost control of his message, there is no running time backward and putting the argument teacup back together, so to speak. But it is tempting to try. Seen in this light, Gould's seemingly sudden advocacy of NOMA in *Rocks of Ages* was not an abrupt about-face but rather an attempt at containment, for Gould had witnessed first-hand the uncontrollability of prophetic dialogue. When the prophet engages in a dialogue with a polity over covenant values, the answer is not always what the prophet wants it to be; that is the political risk of dialogue and why many politicians refuse to engage in it. Instead of "we are a people who value naturalism as a social epistemology," Gould heard louder and louder from his partners in the creation/evolution debate "we are a people who trust our traditional prophets over scientist-prophets."[93]

Through this new lens, let us revisit Gould's *credo* for NOMA:

> I do not see how science and religion could be unified, or even synthesized, under any common scheme or explanation or analysis; but I also do not understand why the two enterprises should experience any conflict. Science tries to document the factual character of the natural world, and to develop theories that coordinate and explain these facts. Religion, on the other hand, operates in the equally important, but utterly different, realm of human purposes, meanings, and values—subjects that the factual domain of science might illuminate, but can never resolve.[94]

This proposed *cordon sanitaire* was not practicable, of course, but in the light of Gould's failed attempts to "resolve" his dialogue with creationists over the previous 20 years, it was an understandable rhetorical response.

Which Channel Broadcasts the Women Scientist-Prophets?

We have now considered the effects of media on the prophetic projects of the Oracles—particularly controversy generation, transformative metaphors, and loss of message control—but a lingering question remains: where are the women in this drama? Mass media science popularization remains strongly gendered toward the masculine—particularly on television, where to this day we find not a single female star of a science-related television program. Women do show up in science programming but typically as assistants or as interviewed experts, not as hosts. They often play a supporting role in written popularizations as well. For instance, a remarkable passage in Louise Sweeney's lengthy profile of Sagan for the *Christian Science Monitor* described Ann Druyan, his wife and collaborator, as "follow[ing] his every word with huge, melting brown eyes in a heart-shaped face. A pretty young woman with long, curling dark hair, she speaks softly and dresses softly: a quiet beige knit outfit." Replace "woman" with "spaniel" in this passage, and Sweeney's message is not significantly altered. Kitty Ferguson likewise made much of how diligently—and literally—Hawking's wife Jane supported him as she attended 24 hours a day to his physical needs. Ferguson quoted Jane's rationale: "I wanted to find some purpose to my existence, and I suppose I found it in the idea of looking after him."[95]

In portrayals like these, the mass media seems to have ignored the potential of Rachel Carson's example and reverted to an Enlightenment ethical paradigm in which only men can be trusted to be truly objective and women are welcome if they keep quiet and wash the glassware. But this observation should not be surprising given the findings that we've reviewed in this chapter, which conclude that the narrative of mass media science is pronouncedly retrograde, masculine, and mythic, favoring the type of the heroic loner scientist/explorer/priest.

Why is television telling such an old story about science? One reason relates to representation. Today there are still fewer women than men in the majority of the sciences and engineering disciplines, particularly in physics and astronomy, which are frequently the subjects of science programs.[96] To host these programs, producers tend to seek out prestigious scientists who communicate well. If fewer of the researchers in a field are women, and if they face retention and promotion obstacles,[97] then the paucity of women stars on television and film is not so surprising. Media scholars have worried for a few decades now that the marginalization of women scientists in mass media is further reinforcing occupational sexism.[98]

The "old story" in science television is not the whole story, however. If we recall the radical voices of Margaret Cavendish, Anna Barbauld, and Rachel

Carson, we remember that women frequently perform the ethos of the kairotic prophet—not just in scientific debates but in social justice movements concerning civil rights (Harriet Beecher Stowe, Sojourner Truth, Rosa Parks, Elizabeth Cady Stanton), welfare (Jane Addams, Mother Theresa), substance abuse (Frances Willard, Candy Lightner), and gender equity (Betty Friedan, Gloria Steinem). Similarly, women have been active and visible in social justice movements related to science and technology. Most media-savvy Americans have probably heard of Erin Brockovich, an activist against occupational poisons, and Jane Goodall, who advocates habitat protection for primates, in addition to Rachel Carson. In each instance these women have voiced the suffering of marginalized agents—human and nonhuman—into the midst of Establishment science. The ethos they performed in these kairoi derived a great deal of power from their stance as outsiders speaking truth to power.

The three Oracles we have studied, by contrast, were not entirely convincing as kairotic prophets. Although they presented themselves as Promethean figures, writing "not for the men in ivory towers, but for us,"[99] and managed to persuade some of their audience along these lines, other readers nonetheless recognized the Oracles as powerful scientific insiders and critiqued their ethos as "arrogant" "hermit[ic]," and "pompous."[100] So, their status as media and disciplinary stars may not have helped them as much as scholars assume. If we compare the impact of the messages of Carson, Brockovich, and Goodall—measured in terms of legislation and regulation—to the Oracles' records in this area, we have legitimate grounds to question the assumption that only the stars of science programs can successfully engage Americans in dialogue leading to certainty and civic action.[101]

Conclusion: Cosmology, Evolution, and Covenant Values

Gould, Sagan, Hawking, and the other late twentieth-century Oracles of science employed mass media to engage lay polities directly in dialogues concerning covenant values relating to the status of science, education, social harmony, religion, and revolution. The Oracles exploited the media's penchant for controversy and its figurative vocabulary to seize public attention. In the process, they discovered that the very tactics they used to cross the traditional disciplinary boundaries between science and popular culture could also serve to reinforce those boundaries, due to the ethical function of Confirmation, which demands demonstrations of privileged access to knowledge to justify prophets' special status.

Why these Oracles? And why during this time? As I mentioned at the beginning of the chapter, previous work has tended to attribute the ascendance of the Oracles largely to their personal charisma. But even an Oracle as self-promoting as Sagan acknowledged that he was just one agent in a larger kairos that seemed to be pushing for greater understanding of the workings and products of twentieth-century science. Regarding this understanding, he surmised,

> In many cases people didn't even know that they hungered for it till they came upon it. And suddenly there's a response. . . . So people personalize it because there's a face and a voice that presents it, but anybody can present that kind of intellectual satisfaction to people every day. So I don't think it's anything deeply personal about me.[102]

In trying to account for the larger kairos that authorized the Oracles, both David Tietge and Thomas Lessl have alluded to a sort of spiritual vacuum left in the wake of Americans abandoning organized religion, a vacuum filled by the Oracles' pronouncements on cosmological and evolutionary origins.[103] If the existence of such a vacuum could be proven, that explanation might well account for the popularity of the Oracles we have considered plus Jacob Bronowski's well-received *Ascent of Man* series and much of *Nova*. But what, then, of Burke's *Connections*, which resolved seemingly incongruous researchers, inventions, and ideas in a murder-mystery-style narrative each week? The series was expected to fail out of the box and instead went on to film two follow-up seasons. And what of James Gleick's best-selling books on chaos theory? It seems more was at play here than a polity seeking its lost origins.

Cosmos reviewer Richard Zoglin brought us closer to the heart of the issue when he explained that "in an era of giant technological strides, when profound questions arise almost daily about where those strides are leading mankind, television is attempting to help satisfy the public's mounting curiosity."[104] The inventions of World War II and postwar science had raised a crop of new moral questions—about the technologies of war, about the relationship among science, government, and industry, about the limits of human intervention in nature and health. These questions seemed impossible to answer without turning to scientists for even technologies to fix the problems. Thus, problematic hybrids of nature and culture proliferated, permanently altering Americans' ideas of both categories.[105] During the ascendance of the Oracles, these problematic hybrids included the ongoing threat of nuclear war, the recent nuclear meltdown at Three Mile Island, and the new possibility of "nuclear winter." People were aware that pollution was increasing; concern was growing over the depletion of the ozone layer. On the home front, the intensifying of the "culture wars" with Ronald Reagan's election to the presidency in 1981 brought the debate over the teaching of evolution in the schools again to a head in a series of state laws requiring "balanced treatment" and the inevitable lawsuits that followed. Homeschooling increased in response to this issue, to failing and underfunded public schools, and to a recession that put a parochial education out of reach for many families. These parents found themselves having to decide what their children should and should not learn in terms of science and religion; in previous generations, these decisions had largely been uncontroversial, delegated to school boards. Now, they were increasingly contentious and even, in some cases, the stuff of national news.

All the issues listed above, including the questions of origins, shared an orientation: they raised questions that scientists supposedly had answers to. But the era of blind trust in scientists (if it had ever in fact reigned) was over; people believed, and media reports reinforced the impression that scientists themselves had contributed to many of these dilemmas.[106] So resolving them required dialogue, not another technocratic fix. Understanding this impetus to dialogue is key to understanding the authorization of the media Oracles.

Just as with Hawking's cosmological chicken and egg, it is impossible to say whether the Oracles or their readers and viewers "started" the dialogue on these dilemmas. But I think we are now in the position to observe that the participants in the dialogue had been assigned their roles long ago—perhaps when Charles II consulted the Royal Society about the Great Plague of London; perhaps even earlier, when Delos asked the Pythia what to do about its own plague. Giberson and Artigas summarized the petitions put to the twentieth-century oracles thusly:

> Are we alone in the universe? Where did we come from? Did the universe have a beginning? Is there a point to our existence? Are we the products of random chance? Where do we find answers to deep and important questions? We are a culture that looks to science because that is where we expect to find our answers. We cannot, however, find these answers ourselves, for only a specialist can navigate the complex terrain that is modern science. We need guides—Oracles—to show us the way.[107]

The dialogue aimed at answering these questions continues. In the early decades of the twenty-first century, it has taken on a new urgency in the debate over climate change. It is to this debate, and to the new prophets of climate science, that we turn in the final case study. In the United Nations IPCC, we find a rhetorically savvy group of scientist-prophets who are familiar with the problems posed by the is/ought divide and by the media and who have worked proactively to remediate their ethos.

9 CLIMATE CHANGE AND THE TECHNOLOGIES OF PROPHECY

"Prophets of doom": that was what S. Frederick Singer, a retired atmospheric physicist, called the IPCC in a recent press release.[1] The quote illustrates two points that should surprise no one at this point in the book: the first is that the most obviously prophetic scientists these days are climate scientists, and the second is that they are engaged in a prophetic battle over who should be authorized to advise the public about future climates, a battle waged with lobbyists, economists, politicians, and even some of their own number—such as Dr. Singer. Singer runs the cheekily titled Nongovernmental International Panel on Climate Change and has developed a reputation for making trouble for scientists who oppose industrial interests.[2]

What it may surprise some readers to learn—as it did me—is how astute the IPCC has become regarding the media construction of its ethos, and how actively and directly it now manages it. These are not the naïve scientists of the 1950s "sweating under the klieg lights" or the wide-eyed media pioneers of the 1980s: these are professional rhetoricians and media producers who exploit the technologies of prophecy—mathematical, verbal, and visual—to mount a sophisticated campaign in favor of their continued authorization as civic oracles.

In this chapter it is crucial to think about kairos and not just documents, and about constitution rather than simply reception. The IPCC's ethos in its Nobel Prize–winning *Fourth Assessment Report* (AR4, 2007) was not a one-off performance but rather a campaign in a prophetic battle that has been two-plus decades in the waging. The "tipping point" in this war came when the IPCC's argumentation on climate change moved to the pivotal stasis of cause/effect. At that point, climate change moved from a scientific dilemma to a political one, and the IPCC began to field ethical attacks predicated on the is/ought stricture—charges of bias and "junk science."

With this background in place, I explore two prophetic technologies the IPCC employed to remediate its fractured scientific-prophetic ethos in the AR4: uncertainty scales and visualizations of climate-modeling results. These technologies are most obvious in the *Summaries for Policymakers* (SPM) that accompanied the report, and the best way I can summarize their ethical effect is via a military metaphor. In the prophetic battle involving the AR4, the IPCC formed a phalanx by (a) decreasing the appearance of uncertainty and disagreement among them, thus closing ranks and making themselves a smaller target; and (b) performing progressive continuity between their disciplinary and prophetic ethē with the help of visualizations of climate models, thus remediating the is/ought chink in their armor.

When we look at the reception evidence for these strategies, it appears the IPCC's rhetorical phalanx was locally and temporarily effective. However, it is important once again to distinguish reception from constitution because when we take a wider view of climate change debates in the United States, it is clear that the schismatic ethos of the climate science adviser persists. I finish the chapter with a brief discussion of Climategate in this light.

The Prophetic Ethos of the IPCC

Climate change is a paradigmatic case of what Funtowicz and Ravetz famously called "post-normal" science—science conducted in a crisis in which both uncertainty and stakes are very high.[3] Such crises call for oracles,[4] and so in 1988 the World Meteorological Organization and United Nations Environment Programme co-authorized the IPCC.

The IPCC is a consortium of thousands of climate scientists, the vast majority of whom represent their UN member nations on a volunteer basis. To date the IPCC has produced four assessment reports on climate change, with a fifth due in 2013. An incredible amount of work goes into these reports—coordinating the efforts of hundreds of lead authors (nearly 500 for the AR4 and reportedly 800 for the AR5), vetting thousands of scientific articles on climate science, polling experts, compiling 2,000-page-plus drafts, and overseeing a complex review process involving external experts and government and industry stakeholders.[5] The IPCC has organized this labor via three working groups, one on the physical science of climate change, one on its socioeconomic and natural impacts, and one on the monitoring and mitigation of greenhouse gases.

On its website the IPCC constructs its organizational ethos as follows:

> The initial task for the IPCC as outlined in the UN General Assembly Resolution 43/53 of 6 December 1988 was to prepare a *comprehensive review* **and** recommendations with respect to the state of knowledge of the science of climate change; social and economic impacts of climate change; possible response strategies and elements for inclusion in a possible future international convention on climate. Today the IPCC's role is also, as defined in Principles Governing IPCC Work, "... to *assess on a comprehensive, objective, open and transparent basis* the scientific, technical and socio-economic information relevant to understanding the scientific basis of risk of human-induced climate change, its potential impacts and options for adaptation and mitigation. IPCC reports should be *neutral with respect to policy,* **although they may need to** deal *objectively* with scientific, technical and socio-economic factors relevant to the application of particular policies."[6] (emphasis added)

We can see reflected in this textual performance the schismatic ethos of the science adviser, fractured by the tug-of-war between the progressive and is/ought models. The disciplinary norms are performed in the *italicized* phrases. Their prophetic ethos as science advisers is indicated by the underlined phrases. The IPCC acknowledges the "is/ought" boundary with the **boldface** phrase "although they may need to"; the progressive model of science-policy integration persists in the boldface "and," which implies that *all* stases fall within the IPCC's original prophetic mandate.

This complicated ethical performance reflects the IPCC's historical experience as a prophetic body. It was founded in response to growing public concern about global warming. Although the concept of "global warming" has been circulating in meteorological circles for more than a hundred years,[7] it blipped onto the public radar in 1980 after a year of record-high temperatures and a report in *Science* that average global temperatures were on the rise. Media outlets around the world followed up on the story. In the United States, NASA climatologist James Hansen, the most unabashedly prophetic of our current climate science advisers, won notoriety with a series of eloquent warnings about the effects of global warming, culminating in his high-profile congressional testimony in 1988, the same year the IPCC was founded.[8] Global warming has remained a political and media fixture ever since, though its coverage has waxed and waned with shifts in both the actual and political weather.[9]

Climate science and climate modeling were in their infancy in the late 1980s, so the IPCC's *First Assessment Report* (FAR), released by the UN in 1990, was somewhat muddled rhetorically as the lead authors struggled to assemble a clear picture of global climate change and its risks.[10] In the United States, these definitional arguments fell comfortably within "risk assessment," which the National Research Council's "Red Book" on risk for federal regulation segregated from the high-stasis activity of "risk management." The Red Book thus institutionalized the is/ought stricture for federal environmental policy making: scientists were to dispassionately assess the state of the environment and calculate its risks, and then this information was to be handed over to politicians, who would calculate the policy.[11] The FAR's adherence to the "is" side of the divide thus placated politicians, who were eager to learn more about climate change after record high temperatures in the 1980s distressed their constituents and confirmed the panel's prophetic call to ascertain the risks.[12]

While the FAR did not seem to expose the IPCC's ethos to critique, their 1995 *Second Assessment Report* (SAR) signaled what Richard Besel has called a "tipping point" in climate change rhetoric: what had been largely a scientific discourse became irreversibly political.[13] An examination of the rhetoric of the report shows that what Besel identified as a rhetorical tipping point was in fact argumentation at the fraught stasis of cause/effect. That the IPCC's rhetoric would shift up-stasis in this way wasn't an unexpected development. Work on the rhetoric of academic disciplines—scientific ones in particular—has suggested that their internal argumentation tends to move up-stasis over time as they grow and entertain ever-more-complex explanations for natural phenomena.[14] This was exactly the situation in the SAR. Whereas the FAR had engaged questions of fact (all historical measurements of temperature show a higher rate of increase since the industrial era began) and definition (this change is anomalous and can be termed "global warming"), the SAR moved for the first time to the cause/effect stasis, arguing that *global warming appeared in part to be anthropogenic*, or caused by human industry. They wrote, "The balance of evidence suggests that there is discernible human influence on global climate."[15]

As soon as the IPCC authors made claims that humans were driving climate change (which in the 1980s was generally thought to be a bad thing, though it had been touted at earlier junctures as a boon that would make more of the earth cultivable[16]), they opened their disciplinary ethos to attack on the grounds that they had transgressed the is/ought boundary by making value judgments—even though they *had not done so*.

Although most of the argument for anthropogenic forcing was limited to the eighth chapter of the SAR, that didn't stop critics from accusing that chapter's lead author, Ben Santer, of pushing an environmental agenda.[17] It did not help that the mid-1990s were relatively mild climatically or that a relatively progressive Executive was attempting to pass emissions legislation against a conservative majority. House Republicans launched the inquiry "Scientific Integrity and the Public Trust" in 1995, targeting the "alarmism" of science advising in the areas of climate change, oxone depletion, and dioxin emissions. They brandished the term "sound science"—meaning high-proof empirical data—as an impossibly high standard for climate projections, in which "probabilities are all we have," in the words of climate science historian Paul Edwards.[18]

The ethical controversy only intensified with the publication of the IPCC *Third Assessment Report* (TAR) in 2001, which made stronger arguments for policies mitigating carbon emissions. Now the attacks came not just from conservative politicians but also from "skeptical" scientists, many of whom were receiving research funding from fossil fuel and other industries that viewed emissions regulation as a threat to their bottom line. These skeptics turned to the Internet, proliferated blogs, and thus gave the public an inflated impression of a full-fledged prophetic battle among climate scientists over the facticity of global warming. The most salient of these skeptical attacks came from the duo of Stephen McIntyre and Ross McKitrick, who worked for the oil-funded Fraser Institute and maintained an active and aggressive critique of the TAR on McIntyre's blog, ClimateAudit.org.[19]

McIntyre and McKitrick singled out Michael Mann, a contributing IPCC author from Pennsylvania State University, and a TAR graph that he had first published with Raymond Bradley and Malcom Hughes in 1998, a graph now immortalized as the "hockey stick." Based on tree-ring data and other sources, it showed a relatively flat temperature gradient throughout human history until the 1950s, when temperatures appeared to spike so steeply they turned the fever chart into an L-shape like a hockey stick. The graph's causal argument was unequivocal—increased CO_2 emissions from industry were causing the earth's surface temperature to rise in an unprecedented fashion. McIntyre and McKitrick expressed doubt about the data and asked Mann to see them. Mann complied but initially kept some back. The Canadian scientists kept after him, and the controversy spread through climate blogs and the mass media until it caught the attention of conservative lawmakers, who convinced the House Committee on Oversight and Investigations to launch an audit of all Mann's federally funded work. He was required to submit all his data, his source code, a complete accounting of all the federal funding he had received, and a detailed "explanation" of his work for the IPCC. His research ground to a near halt until the National Research Council finally exonerated him of all charges of falsifying data in a second investigation in 2006.

Lest we misperceive this controversy as entirely one-sided, the e-mails hacked during Climategate as well as the testimony of bloggers and journalists from this time period reveal that some climate scientists and their supporters were engaging in similar tactics—strategizing how to keep dissenting papers out of scientific journals, putting pressure on journalists to suppress these skeptical views, ramping up the certainty with which particular catastrophic impacts were presented (the unsubstantiated Himalayan glacier melting estimates in the AR4 form the most notorious example of such a jeremiad), and calling climate skeptics "immoral."[20] If it hadn't been before the Mann offensive, the debate was now a veritable prophetic battle with both sides leveraging Internet media—blogs, forums, videos—to vent value judgments and policy recommendations prohibited in the pages of traditional scientific journals.

The pillorying of Michael Mann was very much in the minds' eye of the IPCC authors as they prepared the AR4. The arguments against his ethos had circled around the topoi of the reliability of climate models and the misrepresentation of uncertainty. Accordingly, the IPCC developed specific and novel approaches to these issues in the drafting of the AR4, and it is to those strategies we turn now.

Disciplining Uncertainty in the AR4
Summaries for Policymakers

Uncertainty is *the* rhetorical issue for climate scientists and science advisers. An enormous literature has developed on the topic over the years on technical, epistemic, linguistic, rhetorical, and sociopolitical uncertainty.[21] Within technical uncertainty alone, the category that most intimately concerns climate modelers, there are many "flavors," including error, statistical uncertainty, randomness, variability across the domain to be modeled, temporal uncertainty, structural uncertainty ("unknown unknowns" in the climate situation), metrical uncertainty (mismatches between the model parameters and the real-world situation), translational uncertainty (mismatches between mathematical model results and words used to explain them), and disagreement among experts.[22]

Rhetoricians studying how these kinds of uncertainty are articulated in public debates have discovered that uncertainty is an extremely flexible topos. This flexibility is due to the centrality of uncertainty to public discourse. As Tom Goodnight has argued persuasively, *all* arguments can be understood as attempts to discipline uncertainty—about the world, about a political situation, about the knowledge and attitudes of others.[23] Thus, as we saw in chapter 7, uncertainty can be articulated both to stall action on science-related regulation—"if it ain't broke, don't fix it"—and to push for it via the precautionary principle—"better safe than sorry." Further, uncertainty can serve to both reinforce and break down boundaries between polities. Stephen Zehr documented kairoi in which climate scientists have presented uncertainty as a problem only they can solve. For example, James Hansen recently dismissed the authority of the average citizen to criticize climate predictions: "You can't stick your head out the window to look for the greenhouse effect unless you're clever enough to compare the climate to what it was a few decades ago."[24] This exclusionary stance often underpins requests for funding, as scientists stress their special access to natural knowledge. In funding decisions and similar kairoi, then, uncertainty can serve as an important "boundary discourse" between scientists and politicians—allowing them to work together on something both polities want to see reduced while protecting each polity's autonomy.[25]

This flexibility notwithstanding, scientific uncertainty and political uncertainty are in fact quite different things: one is a fact of the scientific form of life, fundamentally irreducible after a point, whereas the other *can* be functionally eliminated long enough for policy to be pushed through.[26] Contrary to popular belief, political certainty does not automatically follow from a reduction in scientific uncertainty, even under the progressive model. The crisis at hand *must be weighed against extrascientific values* for political certainty to be successfully articulated. This crucial difference notwithstanding, scientific uncertainty and political uncertainty are conflated all the time in science policy making. Both sides are guilty of this conflation. If scientists can express scientific uncertainty as continually diminishing toward political certainty, they can acquire more funding for policy-related research as well as for the basic research that is often higher priority for them.[27] For their part, politicians can flaunt such funding decisions as evidence of progress toward political action; so doing, they cast political issues as scientific issues and push thorny political issues off onto science advisers.[28] In these kairoi the logical divide between scientific and political uncertainty is no easier to maintain than the is/ought divide.

The IPCC has grown into the way they handle and express uncertainty in this complicated context. The term "uncertainty" was never defined in the FAR, and terms such as "certainty" and "expert judgment" were used haphazardly throughout that report.[29] Some of this variability stemmed from the fact that the FAR was, as are all IPCC reports, written by committee. There are at least three tiers of writers for each assessment report—coordinating lead authors, lead authors, and contributing authors.[30] Some working groups poll wider groups of experts as well. In short, there are a lot of cooks, and this has made it difficult to ensure consistent rhetorical practices.

By the time the SAR was being written, it was clear that the IPCC preferred to treat technical uncertainty as a combination of likelihood, confidence, and risk. Likelihood is the percent chance of something happening in the future; the percent is usually arrived at by averaging expert estimates. The "spread" in these expert estimates is the confidence interval: so, if a group of experts reports a 60% likelihood of a Category 5 hurricane with a confidence interval of ±4, that means 95% of the experts gave an estimate somewhere between 56% and 64%. The smaller the confidence interval, ironically, the more confident the reader can be that the experts have reached consensus. Risk, finally, is the likelihood of a future event multiplied by its potential impact—for example, the hypothetical hurricane's category rating times the damage to persons and property it would do. Expressing risk for very unlikely but extremely impactful events has proven a challenge for the IPCC authors.[31]

The SAR tried to clarify these key terms, but again, the committee drafting process created consistency problems. Confidence intervals were given as ranges, as explained above, but likelihood was not quantified. The SAR SPM did not address uncertainty at all.[32] Especially after the furor surrounding the strength of claims for anthropogenic warming, the IPCC realized this oversight was a costly one. The IPCC's "uncertainty police," Richard Moss and Stephen Schneider, warned lead writers in advance of the TAR that policymakers would need to take action long before scientific uncertainty could be reduced to everyone's satisfaction. It was the IPCC's job to present future scenarios as clearly as possible so policymakers could make informed decisions. Therefore, it was important to avoid statements of "medium confidence" because these were very difficult to translate to a policy recommendation.[33] To further help policymakers evaluate the reports, authors were encouraged to address structural and metrical uncertainty—issues related to how accurate and comprehensive the climate models were. Ideally, authors would provide a "traceable account" of how they arrived at their estimates. In particular, they should be forthcoming about problems with averaging expert estimates, with special attention to split decisions and outliers.

To strengthen scientific consensus, Moss and Schneider recommended an expert elicitation procedure called (I am not making this up) the Delphi Method. Each expert is asked for her likelihood estimate for some future event; then, all the experts see a probability distribution function of all the (anonymous) predictions; each expert is then asked to revise her prediction. This process usually moves all the estimates toward the mean, tightening the confidence interval.[34]

Moss and Schneider's recommendations had a marked effect on the TAR, published in 2001. This report introduced for the first time a confidence scale, and ranges were usually (but not always) given with 95% confidence intervals. The report also explained that likelihoods were calculated using Bayesian probability—a subjective assessment based both on statistics and expertise. Named for the Enlightenment mathematician Thomas Bayes, the Bayesian likelihood of a proposition like "global temperatures will warm 2°C on average by 2030" constitutes "the odds that an informed better would accept that [the statement] is true."[35] Such an estimate factors in many forms of uncertainty that purely frequentist likelihoods cannot, such as expert assessments of the reliability of the models used (structural and metrical uncertainty).

But in spite of the improvements, the three working groups still did not treat uncertainty in the same way, and a reception study found that readers interpreted verbal expressions of uncertainty such as "highly probable" in idiosyncratic ways that were not much affected by the definitions provided in the TAR.[36] Meanwhile, media reports had been amplifying the public perception of lack of consensus among climate scientists.[37] Uncertainty became a primary concern for the IPCC authors going into the AR4. They developed a two-pronged strategy with respect to expressions of uncertainty: first, discipline it to make it appear manageable and reducing; second, emphasize certain types of risks. The result of these efforts was the IPCC's most certain-sounding document to date.

Two internal documents laid out the new strategy: the Manning report and the *Guidance Notes for Lead Authors*. The Manning report, which had presented the results of the reception study, stressed that confidence intervals must be tight in the AR4. It recommended that any likelihoods with wide or bimodal ("split decision") confidence distributions should be left out of the report altogether unless they related to what came to be called "reasons for concern"—low-probability but high-impact events, such as hurricane swarms or widespread pollution of drinking water. Some rhetorical advisers among the IPCC authors felt that these tipping-point-type impacts had be downplayed in previous reports at the cost of political certainty and action.[38]

The official IPCC *Guidance Notes for Lead Authors* was a brief document synthesizing the Manning recommendations with Moss and Schneider's paper. It presented new likelihood and confidence scales that all lead authors were to employ to standardize the treatment of uncertainty across Working Group reports. These scales divided likelihood and confidence into percentiles and assigned each one a verbal translation, ranging from "extremely unlikely" to "virtually certain" for likelihoods, and "very low confidence" to "very high confidence" to represent consensus. Writers were cautioned to stay away from the two bottom tiers of the confidence scale except in the case of "areas of major concern"; likewise, the *Notes* encouraged authors to pay "special attention" to risks of "high consequence."[39] Finally, the *Notes* recommended using visual representations of uncertainty where possible, for reasons we will see shortly.

In contrast to the recommendations of the Manning report, which were almost all aimed toward reducing the appearance of uncertainty and stressing key risks, the *Guidance Notes* contained some more traditional cautions aimed toward the performance of disciplinary ethos. It warned authors not to "trivialize" the expression of future scenarios to get a higher confidence rating from experts; to distinguish topics on which consensus should be sought from those on which a range of views should be presented; to guard against "cascading uncertainties" by detailing the types of uncertainty that underlay each decision input from model building to policy recommendations; and, finally, to avoid "value-laden" language when discussing impacts.[40] These pieces of advice appeared to contradict other admonitions toward certainty and risk language and for the most part were disregarded in the *Summaries for Policymakers,* with a few exceptions.

First, in defiance of recommendations to elaborate sources of uncertainty, the SPM authors gave no account of sources and types of uncertainty underlying their projections, and they cast scientific uncertainty as constantly diminishing as a result of research; by doing so, they implied it might be possible to attain certainty about climate change in the future—exactly what the polities who consult them wanted to hear.[41] The *Synthesis Report* (SYR) SPM reiterated the myth of the constant progress of science with phrasings like these: "There is now higher confidence than in the TAR in projected patterns of warming"; and "Studies since the TAR have enabled more systematic understanding of the timing and magnitude of impacts"[42] These statements implied that once scientists knew more certainly what would happen and how serious it would be, the decision-making process would become easier for policymakers.

In spite of this language tending toward certainty, the Working Group I (WGI) SPM authors did appear to heed suggestions to avoid value-laden language. The introductory paragraph of the SPM demonstrates this attempt perspicuously:

> The Working Group I contribution to the IPCC Fourth Assessment Report describes progress in understanding of the human and natural drivers of climate change, observed climate change, climate processes and attribution, and estimates of projected future climate change. It builds upon past IPCC assessments and incorporates new findings from the past six years of research. Scientific progress since the Third Assessment Report (TAR) is based upon large amounts of new

> and more comprehensive data, more sophisticated analyses of data, improvements in understanding of processes and their simulation in models and more extensive exploration of uncertainty ranges.[43]

Words such as "data" and "findings" invoke the stasis of fact; verbals such as "understanding" indicate arguments about definition; and language like "processes and attribution," "drivers," "change," and "analyses" targets the key stasis of cause/effect. Disciplinary ethos is further performed by hedges demonstrating humility ("estimates," "projected") as well as by arguments for the objectivity and rigor of the findings in the SPM. *Nowhere* here do we find explicit evaluations (good/bad/dangerous/etc.) or recommendations. The closest the language comes to these stases is the term "assessment," but it refers to past IPCC efforts and cannot be taken for prophetic argument.

This rhetorical asceticism, however, did not hold throughout the SPMs. The likelihood and confidence scales injected evaluation into the very core of the arguments. Phrases such as "likely" are subjective evaluations, even though they index technical values like probability and efficiency. While such technical evaluations can pass for low-stasis claims, they are not; this becomes manifest when you pair an evaluation such as "likely" with an impact that is clearly bad, such as "intense cyclone activity."[44] This is the language of risk, which, just as I argued for Rachel Carson's case, lends itself to interpretation as prophecy.

And so, in spite of the attempts to avoid value-laden language and to concretize uncertainty, the AR4 wound up the most certain-sounding prophetic document the IPCC had produced. The first line of AR4 SYR SPM was "Warming of the climate system is unequivocal." This SPM, which addressed some of the same arguments as the earlier WG1 SPM, contained *not a single evaluation* of "low confidence" for any likelihood presented. "Medium confidence" was the lowest level used on the scale, and only in reference to "areas of major concern" such as infectious disease vectors. Most importantly, the likelihood and confidence scales were absent from *all* the SPMs, which strengthened the impression of scientific consensus and confidence on climate change: after all, "High Confidence" suggests a totality that "about 8 out of 10 chance…of being correct" does not.

Further, the AR4 SPMs' focus on "reasons for concern" (RFC) amplified risk, thus increasing the prophetic tone of the document. The RFC were introduced in the TAR: loss of unique/threatened ecosystems, extreme weather events, disproportionate impacts on the poor, compound effects on markets, and "large-scale singularities" like permanent coastal flooding. They were the foundation of the controversial "burning embers" graphic in the TAR, which showed the risk of each RFC as a bar grading from white to orange to red as risk increased. Although several key writers for the IPCC wanted to include an updated version of this graphic in the AR4—one that was significantly "hotter"—four reviewing nations (the United States, Russia, Saudi Arabia, and China) voted it down on the grounds that it expressed too much "judgment."[45] The AR4 kept verbal descriptions and likelihoods for the RFC, however, defending them as follows: "The five 'reasons for concern' identified in the TAR remain a viable framework to consider key vulnerabilities. These 'reasons' are assessed here to be stronger than in the TAR. Many risks are identified with higher confidence. Some risks are projected to be larger or to occur at lower increases in temperature."[46] This statement was accompanied by a tabular version of the burning embers graph (SYR SPM-7), which replaced the heat/risk appeal of the red bars with cool pastel backgrounds and verbal expressions of impacts: "hundreds of millions...exposed to drought or water stress," "up to 30% of species at increased risk of extinction," and "substantial burden on health services." While these jeremiads might seem restrained compared to the doom-mongering we saw in classical prophecy, the IPCC's warning nonetheless chimed out loud and clear here: turn or burn, quite literally.[47]

These judgments, and the prophetic stance from which they were issued, were recognized readily by audiences both supportive and critical. In pro-conservation media, the AR4 was touted as the coup de grace for skeptics who had insisted there was no scientific consensus on anthropogenic warming.[48] The support groups behind these messages—the Sierra Club, Environmental Defense Fund, Natural Resources Defense Council, and so on—were critical to the constitution of the IPCC's prophetic ethos. They repeated and amplified the AR4 findings in the blogosphere and news media because they heard the IPCC espousing their covenant values. These politics of recognition remind us that prophecy is not a monologue by the prophet but must be constituted via political dialogue with supporters.

Among these supporters, a national phone poll by Stanford University found an increase in confidence that global warming was real and anthropogenic after the publication of the AR4.[49] More intriguing was a sharp spike in the use of the phrase "global warming denier" immediately following the publication of the WGI SPM. Google Timeline, which graphs the occurrence of search terms on news sites and blogs over time, shows that instances of the phrase nearly *doubled* between January 2007 (98 uses) and February 2007 (182 uses). Since that month, the term has remained in higher overall currency in Internet news than it has at any previous epoch in Web history. "Denier" is a much stronger term than "skeptic," implying willful disbelief. It has been applied to those who deny the existence of the Holocaust and other atrocities. Using this epithet, therefore, is a serious ethical risk for news writers, who are in general are supposed to avoid ad hominem attacks, even in op-ed pieces. If twice as many news sites were willing to take this risk after the publication of the SPM, it stands to reason that they felt licensed to do so by the scientific consensus projected in the AR4.

Even opponents appeared to give ground on this point. They recognized the consensus articulated in the AR4, the certainty it appeared to project, and moved immediately to criticize the IPCC's prophetic stance. The *Washington Pest* tartly commented, "Anyone who doubts that the new IPCC *Summary for Policymakers* is an advocacy document is ineligible for duty on the jury of reason."[50] The "IPCC Criticism of the Week" blog called out "biased" statements in the SPM.[51] Indur Goklany, writing on the Cato Institute's blog in September 2008, accused the IPCC authors of "perpetrating a fraud on the readers" in the Working Group II SPM by maintaining a doomsday focus on drought when global warming in fact would cause some areas to receive above-average rainfall.[52] And so on.[53] The IPCC's jeremiads were predictably leveraged as grounds for attacks on their disciplinary ethos—the inevitable opportunity cost for science advisers of choosing a strong prophetic stance. But critiques revolving around the old issues of consensus and uncertainty were practically nonexistent in the Internet response to the AR4. The IPCC's attempts to discipline uncertainty appeared to have worked, at least temporarily. Audiences, supportive and critical alike, apparently agreed to recognize the IPCC as they presented themselves in the SPMs—as a unified oracle calling for change in the face of certain global warming.

Visualizations in the AR4 SPM: Remediating Ethos

If the IPCC's verbal strategies aimed at closing ranks and presenting a unified front, their visual strategies sought to mediate ethical attacks by covering the vulnerable gap between the IPCC's disciplinary and prophetic ethē. The most dramatic rhetorical devices in the SPM, the ones that have been copied and posted on blogs throughout the cybersphere, are the report's graphics, most of which represent the output of climate models.

Computer models, particularly global circulation models (GCMs), are climate science advisers' primary technology of divination. GCMs slice the earth's surface and low atmosphere (about 10 kilometers of it) into cubes. An array of data relating to climate circulation populates each cube: surface temperatures, air pressure, and water vapor are the most common data sets. The cubes become "points" that are combined in a grid, and algorithms are written to govern how the grid changes over time given the starting conditions and certain assumptions—say, CO_2 increasing at an estimated rate.

Processes that occur entirely within a cube—such as cloud formation—are invisible to the model, so they are "parameterized," which means that they are ignored, simplified so the model can handle them (e.g., clouds are modeled as a thin layer of water vapor across the cube at a certain height), or micro-modeled and then input to the main GCM.

With these data, rules, and parameters in place, the GCM is started and stopped to see how it changes over time. The first runs go backward, "predicting" the past so that the model end states can be compared against historical data. If anomalies occur (like the oceans turning into giant skating rinks—which actually happened in some early GCMs), flux corrections force the model to stay within known limits. When the model is corresponding well to recorded global data (a 75% match is considered "well"), the model is started in the present and allowed to run into the future.[54]

This is how climate scientists predict the future. As Stephen Schneider put it, "All we have is theoretical evidence and modeling evidence now. The way we test our models is we run them on historic cases, and when they give us some semblance of reality in historic cases then we at least have modest confidence in them to project the future."[55]

Obviously, the output of the GCMs depends a great deal on the assumptions, algorithms, and parameters input to them.[56] These are crafted based on the best guesses of climate scientists. And they are hard to validate since most GCMs tend to use the same sets of assumptions. It's a little like checking one clock in your house against another one to make sure it's not running slow—how do you know they're not both running slow compared to Greenwich Mean Time? Another problem is that many of the data input to these models come from other models, like the micro-models of cloud behavior. Technical uncertainties multiply in ways that are hard to control.

These problems with starting assumptions and uncertainties are the central topoi of criticism of GCMs. They have consistently served as rhetorical grounds for attacks on the ethos of the models' programmers: since they selected the assumptions, the ethical charge goes, those assumptions are infected with the programmers' presumed leftist politics.[57] To defray such criticisms, modelers have been advised to state their assumptions up front.[58] The reality for the IPCC, however, is that space is at a premium in the SPMs; further, if politicians and lay readers can't understand how GCMs work, what are the odds that they'll understand and approve the modelers' explanations of their starting assumptions? Such attempts to discipline metrical uncertainty can ironically serve to amplify it in readers' perceptions.[59] And amplifying uncertainty tends to disintegrate the ethos of those called upon to ascertain the future.[60]

The IPCC authors were well aware of the ways in which climate models had exacerbated the fractures in their ethos when they wrote the AR4. Martin Parry, a WGI SPM author, attempted to justify the report's climate models in a BBC radio interview:

> Orthodox science has difficulty predicting the future, especially if we have experienced nothing like it in the past. Computer models are essential to these predictions. But to many non-scientists, they are an unknown quantity. However, a new development in the Fourth Assessment is that it concludes, from an examination of 29,000 data sets, that the impacts of climate change occurring now can be observed everywhere on our planet…This is traditional science-based observation and measurement, not 'arm-waving' with computer models.[61]

With the phrase "traditional science-based observation and measurement," we see Parry confirming the IPCC's prophetic ethos via its disciplinary ethos. He treats the measurements reported in the AR4 as confirming both the TAR's predictions and the IPCC's ethos as a civic oracle. To paraphrase: as scientists, we are now measuring the negative effects that, as prophets, we warned you were imminent.

Parry's framing of the GCMs as key to the IPCC's ethos suggests that the models can be an ethos booster as well as a liability.[62] Models can lend this kind of support because they can actually function as ethical proxies in public debates about science. Bruno Latour's actor network theory makes a place in the network for graphics like climate model visualizations as "mediators," agents that pass messages between the nonhuman world and the other human agents in the network. While mediators can multiply and change messages unpredictably, crucially in the message there is some "immutable mobile" that is preserved, the thing that makes the mediators useful for scientists.[63] For climate models, the "immutable mobile" is the cause/effect relationship between carbon emissions and surface temperature.

To translate Latour into the rhetorical terms we have been using, climate model visualizations, even though they are not people, can have agency in a kairos involving scientific debate. Because they can have agency, they can have ethos since ethos is not a personal quality but an expected performance constituted by audiences. Carolyn Miller arrived at a similar conclusion in her studies of two separate science-advising kairoi, one involving the Rasmussen Report on nuclear safety and the other, expert systems in artificial intelligence research. In both situations, she observed science advisers employing their calculations and models as forms of "rational ethos." The implicit argument ran something like, "We're not the ones making the predictions, the models are; we're just faithfully reporting the results."[64] Lawrence Prelli has similarly documented the use of maps as ethical proxies in a territorial dispute between US and Canadian science advisers, during which the Canadian adviser was compelled to refute the US's depiction of "natural boundaries" in the water despite his stated belief that no such thing existed. In other words, Canada found they had to go through rather than around the US maps to "reach" the American position.[65] In all three of these cases, technical models served as proxies for the science advisors' ethē in public debate.

The IPCC authors chose a similar strategy in the AR4 SPMs. Roughly equal space in the WGI SPM is dedicated to graphics and words, and 70% of the graphics are visualizations of the results of computer models. All these visualizations exhibit one of both of the following characteristics: time series proceeding from past to future (left to right) showing increasing trends; and color-coding using red to indicate future undesirable warming directly or indirectly (as in the concentration of greenhouse gases). Each of these visual strategies—time series and color-coding—served as ethical proxies, or "mediators," in Bruno Latour's terms.

First, the time series: the most famous graphic in the AR4 is figure WGI SPM-5 (a.k.a. SYR SPM-5), versions of which appeared on over a dozen websites and blogs during 2007 and 2008, and for good reason—it is a beautiful and terrifying graphic (see color plate, figure 8). The vertical line at the center indexes the present (roughly): the observations to the left of center fall in the past and are colored in somber, definitive black; the brightly colored lines to the right represent possible future temperature trends depending on carbon emissions scenarios. Of these, the most extreme is colored red and dominates the visual impression of the graph because of both its brightness and its topmost position as it rockets up toward a predicted 3.6°C increase in global average temperature by 2100.

SPM-5 is a relatively novel graphic for the IPCC. In the TAR SPM's, the majority of climate visualizations treated either the past *or* the future. The only graphics that integrated past with future presented historical climate records as spotty or thin lines relative to a wide blur of variability; future predictions were either dotted lines muted by a gray "error" field or an indistinguishable array squashed into the last tenth of the time series. In stark contrast to these choices stands the equal time that AR4 SPM-5 gives to past and future as well as the bold width of its trend lines—roughly 30% thicker than those of its TAR predecessors—and their bright color values. The result is a visual de-emphasis of technical uncertainties, even if statistically speaking they were equal to or greater than the uncertainties in the TAR visualizations.[66]

Understanding the significance of these changes in strategy requires a vocabulary for understanding the rhetoric of images. In *Reading Images: The Grammar of Visual Design*, Kress and Van Leeuwen identify three key vectors that control information in graphs like the SPM 5: given—new, which tends to run from left to right in deference to the tendency of European languages to place known information at the beginning of sentences and new information toward the end; ideal—real, which orders these philosophical arguments left–right or top–down; and center-margins, which places the most important element at the center and its causes or consequences at the margins. When these dimensions interact in a single graphic, you get a triptych in which a starting state on the left and an ending state on the right are connected through a mediator. The most common triptychs are religious, in which a sinner on the left is transformed into a saint on the right through the mediation of Christ, or a virgin is transformed into a Madonna through the mediation of a central angel.[67]

SPM-5 is a triptych. Even though it appears to have only two parts, past and future, we cannot overlook the strong central emphasis on the present as indicated by the vertical black line and by the shift from black-and-white to color values. The present is the triptych's central panel or mediator.

Now for the use of color in the SPM-5: Kress and Van Leeuwen argue that the use of color expresses modality; it communicates the artist's confidence in the truth of what she is presenting. For scientific graphics, black-and-white expresses a strong argument for the truthfulness of the imaged message; color expresses speculation.[68]

If we use Kress and Van Leeuwen's color theory and map their concept of the mediator onto Latour's mediator, we have a compelling account of how SPM-5 represented the IPCC's ethos by proxy. The historical past, represented in the triptych by the bold black line, is the traditional rhetorical turf of the IPCC's disciplinary ethos, where the "observations and measurements" it makes regarding facts, definitions, and causes/effects are considered true and reliable. The brilliant branching futures of SPM-5 signal the prophetic ethos of the IPCC; after all, the whole reason the IPCC was called in the first place was to project future climates so polities could take action. The center of the triptych is the tipping point—the time when the models are started, when the projections are made. As I argued earlier, the tipping point in climate debates has been the stasis of cause/effect. And indeed, cause/effect is the rhetorical engine that drives the GCMs: causal assumptions turn present conditions into likely future effects.

It is this crucial center panel of the triptych that "remediates" the fracture in the IPCC's ethos by covering it with a performance of continuity and certainty. The visual impression of the graphic expresses confidence in both the historical measurements and the projections. As Gilbert and Mulkay have observed, scientific visualizations tend to be interpreted by nonscientific audiences as factual representations of reality.[69] Given this tendency to strengthen the illusion of certainty as well as their ability to carry ethos in public debates about science, graphics such as SPM-5 are well positioned to make ethical cases for their authors. SPM-5's ethical defense can be paraphrased, "See? Just as you trust us to measure the past, you can trust us to predict the future; and just as you trust our predictions, you can trust our evaluations and recommendations." As Hooke's microscopic engravings confirmed the Royal Society's privileged access to natural knowledge, and as the detailed graphics of nebulae and evolutionary taxonomies came to stand in for reality in the media arguments of the 1980s popularizers, so SPM-5 stands in both for the future climate and for the privileged scientific-prophetic ethos of the IPCC.

If it seems that I am overloading SPM-5 with ethical arguments that its authors never intended and its readers did not engage, the wider constitution context validates my reading. In terms of the IPCC's intentions, Parry's argument quoted earlier explicitly links climate models to the group's ethos. This link was corroborated in an oral presentation of the WGI SPM that I attended in 2008, during which IPCC author John Overpeck repeatedly directed attention away from himself and onto the visualizations he presented on slides while he predicted future warming in the southwestern United States. Further, he repeatedly emphasized the temporal and ethical continuity of SPM-5 and related graphics by sweeping his laser pointer from left to right, past to future, across the graph. So, both my proxy and continuity arguments about the graphics were borne out in these IPCC performances.

In terms of reception of SPM-5 as an attempt at ethical remediation, there is some evidence it was so received by the blog authors who reproduced it. In the 14 reproductions of SPM-5 I found from 2007 and 2008, the graphic triggered only arguments about its projections and parameters; in no instance did it facilitate an ethical critique of the IPCC, even for blog authors who were critical of the IPCC in other posts. Just as Prelli and Miller found in their studies, SPM-5 appeared to succeed in presenting the IPCC's ethos as a function of its technical aspects.

As the IPCC scientists contributed to the constitution of their ethos in blog arguments following the publication of the AR4, they continued to treat the model visualizations as proxies for their credibility as climate scientists. In some cases they defended the models; in others they shed them like damaged

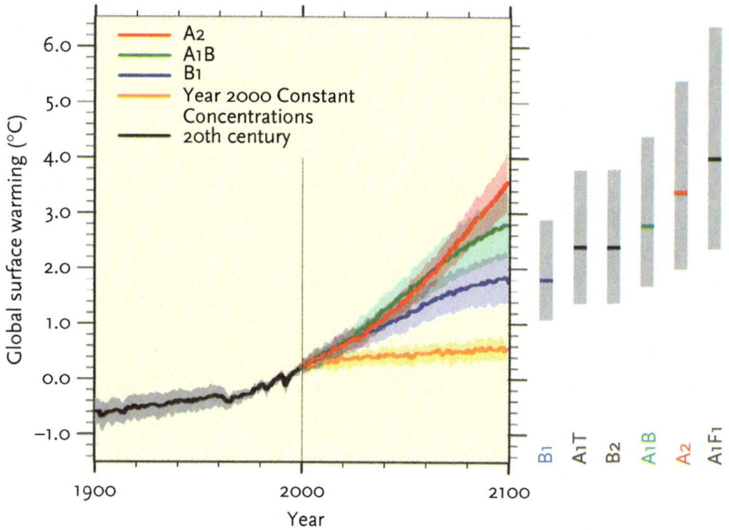

FIGURE 8 WGI SPM-5 from the Intergovernmental Panel on Climate Change's (IPCC) *Fourth Assessment Report* (AR4, 2007) published by Cambridge University Press. I argue this figure functions as an ethical mediator: via the predictive power of climate models, it performs the continuity of the IPCC's disciplinary ethos (represented by the solid black line of historical temperature data) with its prophetic ethos (the bright branching futures).

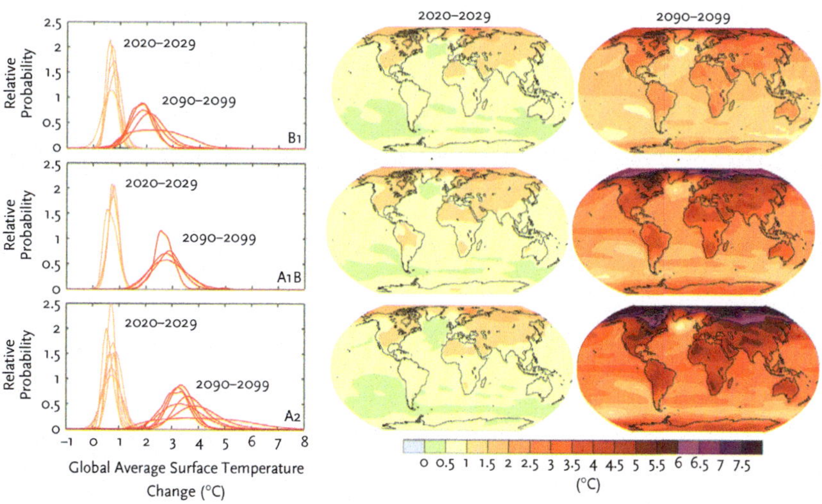

FIGURE 9 WGI SPM-6 from the Intergovernmental Panel on Climate Change's (IPCC) *Fourth Assessment Report* (AR4, 2007) published by Cambridge University Press. These maps represent on a global scale some of the warming scenarios depicted in SPM-5. The naturalistic shading of the reds encourages viewers to see these as realistic rather than speculative projections.

pieces of armor, drawing a distinction between these errors and the overall legitimacy of the IPCC. This latter strategy scaffolded a protracted debate between Gavin Schmidt of NASA and Roger Pielke Jr. on Pielke's science policy blog, *Prometheus*.[70] Such model-shedding maneuvers have been found to succeed elsewhere in diverting criticism from scientists' ethos and perpetuating public support for their research.[71]

Most other graphics in the WGI SPM followed the argumentative pattern of SPM-5, but a few, such as SPM-6, added maps to help the users orient themselves in the future climate; that is, they could find their homes and see how hot they were projected to get (see color plate, figure 9). The color red was again used to signify warming, but the usual code equating color to speculation in scientific graphics was altered somewhat by naturalistic shading of these reds, imitating the way we see heat spread and fade in a sunburn or campfire embers. Naturalism or realism in images invites viewers to take the presented argument as fact rather than speculation.[72] This appeal was intensified by the topos of the map itself, which encouraged viewers to think they could comprehend the entire future climate situation.[73] These appeals to realism and comprehension, coupled with the alarming reds and the invitation to locate themselves on the map, encouraged viewers to project themselves into the frightening future—just as Jeremiah's vivid descriptions of plagues and wars to come invited his listeners to reflect on how they would feel during such disasters.

How does a visual jeremiad such as SPM-6 interact with the IPCC's ethos? As we saw in classical prophecy and also with Rachel Carson's use of the formula, the jeremiad's indictment taps into the audience's repressed guilt for having transgressed covenant values, while its vividness confirms the prophet's privileged access to knowledge. The fear that the jeremiad mongers casts the audience back on the prophet as their only hope for more information or intercession with the divine judge. So, jeremiads galvanize emotional responses to crises while simultaneously bolstering the ethos of the prophet who achieved the dystopic vision.

Did the users of SPM-6 react to its jeremiad in these ways? I have several sources of constitution evidence that seem to indicate they did. At Overpeck's talk, an audible gasp accompanied his clicking over to a graph related to Figure 9, which showed the surrounding region rather than the whole world to facilitate the audience's self-location on the map. Recall also the strongly negative response of some contributing nations to the "burning embers" graph. In addition, responses to the WGI SPM on critical blogs predictably accused the IPCC of "alarmism," "hype," "exaggeration," and other terms associated with the jeremiad mode.[74] But finally and perhaps most tellingly, some of the IPCC's key prophetic opponents adopted the same color codes in their responses to the WGI SPM. The Fraser Institute's "Independent Summary for Policymakers" employed the same overall color code throughout the document;[75] tellingly, however, they foregrounded the "cool" colors of blue and black in their fever charts to reinforce their argument *against* the existence of anthropogenic warming. The Fraser's replication and discussion of so many of the IPCC's graphics indicates that the IPCC authors were at least partially successful in presenting these graphics as ethical mediators; McIntyre and McKitrick chose to attack the models instead of launching another ethical smear campaign as they had against Mann after the TAR. In each case, the IPCC positioned itself as the principal diviner of climate futures.

The Constitution of the IPCC's Ethos Surrounding the AR4

To summarize the strategies the AR4 authors used to constitute their ethos in the ongoing prophetic battle over climate change: they formed a rhetorical phalanx by closing ranks and raising shields. The authors disciplined uncertainty and emphasized risks to increase the appearance of consensus and scientific certainty in the AR4, thus limiting those grounds for ethical attacks. In terms of the visual strategies, we see an emphasis throughout the SPMs on linear trends that performed the integrity of the authors' scientific-prophetic ethos. These elaborate metaphors based on divinatory technologies served as ethical mediators that bridged the key vulnerability in the IPCC's ethos—the is/ought schism. Further, the use of graphic appeals to "hot" and "home" invited audiences to convert their fears into reliance on climate scientists as their best options for certainty and salvation. Thus, the AR4 strategies, taken together, succeeded *locally and provisionally* in bolstering the IPCC's authorization as a civic oracle for divining future climate effects and policy recommendations.

I say "locally" because visual rhetoric in science is much more contingent than verbal rhetoric, meaning that how audiences interpret graphics depends more on context and is therefore less predictable than how they interpret written claims about science.[76] Thus, while we might expect other science advisers—economists or seismologists, for instance—to discipline uncertainty claims the way the IPCC did, we would not expect to suddenly see "triptychs" or "burning worlds" proliferate in their reports. These were contingent strategies shaped by the particular ethical crisis the IPCC faced and the particular resources—human, technical, and otherwise—that they had at their disposal while crafting the AR4.

I say "provisionally" because we must remember that ethical attacks against the IPCC persisted in the wider contexts of climate change debates, as I noted at the end of the uncertainty section earlier. It is tempting to cite the Nobel Prize and the jump in American climate-science funding that followed it as evidence that the AR4 succeeded in remediating the IPCC's ethos on a more global scale. However, it is much more likely that the Nobel Prize, the efforts at performing certainty in the AR4, and the funding bump were all responding to and perpetuating a wider political crisis over global warming. While national polls such as those conducted by broadcast TV news organizations and by Gallup are notoriously untrustworthy, social scientist Jon Krosnick argued in 2010 that the better-executed phone polls (probability sampled and asking simple questions like, "Do you think global warming exists?") reveal a consistently high level of belief in global warming among Americans despite recent claims that this belief is decreasing. According to Krosnick's results, a staggering 82% of Americans expressed belief in and concern about global warming the year before the AR4 was published.[77] Any bid the IPCC has made to change its ethos has to be considered in the wider kairos of this debate, which must include an account of Climategate.

Climategate

In November 2009, a dataset hacked from the Climate Research Unit at East Anglia University—a premiere climate institute and the institutional home of several IPCC lead authors—cast doubt on the consensus the IPCC had performed in the AR4. The hack, christened "Climategate" by skeptical media outlets, contained emails among IPCC authors dishing internal squabbles, ridiculing the opposition, and debating "tricks" for representing climate change data persuasively.[78]

Climategate sparked a flurry of criticism accusing the scientists of having "hidden" disagreements and uncertainties rather than having diminished them.[79] Critics devoted particular attention to old e-mails from climate scientists expressing doubts about Mann's methods in constructing the "hockey stick" graph, methods that these same doubters would later fight vehemently in public to defend. This discovery rekindled debates on *Climate Audit* and other skeptical climate blogs about the validity of the TAR and AR4 and the trustworthiness of their authors.

The IPCC authors largely avoided reentering these debates. If they responded at all, they attempted to quarantine their scientific ethos from their prophetic policy-making ethos. Gavin Schmidt fessed up to writing some of the nastier caricatures of IPCC critics found in the cache but then reminded readers: "Newton may have been an ass, but the theory of gravity still works."[80] Predictably, these types of quarantine arguments appeared to be untenable with critics. Most of the attacks used any involvement in politics by climate scientists as grounds for discrediting their scientific findings. For example, *Finem Respice* accused IPCC scientists of acting for political gain:

> As a group one rarely sees scientists (or, indeed, any vocational group other than politicians) so deeply in love with the by-hook-or-by-crook of politics, the grand import of jetting off to Nice for the next climate meeting and the limelight that accompanies all these world-saving goings on as those few, those lucky few exposed in the CRU emails . . . It is all but impossible not to come away with a sense of what is plainly a naked lust for naked ambition simply oozing out of those texts. I am utterly devoid of sympathy for any such that later claim to have been forced to compromise their composure, their decorum or their data because of the unfortunate realities of politics.[81]

Similarly, on his *Climate Skeptic* blog, Warren Meyer insinuated that bias was the only explanation for one TAR author's failure to publish his private concerns (expressed in one of the hacked emails) about the "hockey stick" graph: "When we understand the incentives that are driving him to suppress his own scientific views, and to publicly ridicule those who share his private concerns, we will understand better what is broken in the climate science process."[82]

What Climategate teaches us is that an ethical meme that has developed over hundreds of years is not likely to mutate radically in one performance. Although the IPCC may have succeeded in convincing its audiences to recognize the integrated ethos it presented in the AR4, such an acceptance was provisional and did not significantly alter the ethical force of the is/ought stricture. The Climategate attacks followed the pattern we have seen develop since the seventeenth century: any hint of "incentives" or "interest" in the outcome of a scientific question disqualifies a scientist-prophet from being able to effectively channel natural truths on behalf of the public. That these old arguments still take place in an era when everyone knows that the IPCC is convened by a political organization and that the Nobel Prize is an activism prize, attests to the incredible staying power of the current meme of scientific-prophetic ethos. The results are therefore as predictable as the attacks: when climate scientists make predictions and policy recommendations at the behest of government agencies, the factions whose politics align with these pronouncements laud the trustworthiness of the source; those whose politics are out of phase with the pronouncements accuse the scientists of "junk science," bias, and corruption.[83] Policy founders in the resulting impasse over the trustworthiness of the experts who have been petitioned to resolve a dilemma that policymakers felt was beyond their ken in the first place.

And yet any meme can change under the right amount of selective pressure. Thus, the provisional ethical victory the IPCC won in the AR4 may signal the beginning of a wider and more lasting mutation in the way the ethos of the science adviser is constructed in the United States. I evaluate the odds of such an ethical shift in the concluding chapter of this book.

10 POSTLUDE

PROBLEMS AND SOLUTIONS

On April 6, 2009, a magnitude 6.3 earthquake struck the idyllic Italian town of L'Aquila, killing 309 people and displacing 65,000. Just a week before, a government commission on civic safety in the region held a press conference at which they had reassured residents of the very small likelihood of a serious earthquake. This announcement had been in part designed to obliquely counter a series of alarming (and inaccurate) earthquake predictions made by a local radon-detector enthusiast. But after the devastation, L'Aquila residents lashed out at the commission for the appeasing messages it had broadcast. In an unprecedented move, manslaughter charges were filed against one commission member and six advisory seismologists. They were convicted of these charges in October 2012 and sentenced to six years each in prison, a verdict they are currently appealing.

The seismologists have protested that they were only responsible for advising the commission on earthquake likelihoods and risks and that it was the commission's job to convert these scientific uncertainties to political certainties in the form of hazard assessments and public safety information; they, and the thousands of scientists who have written letters to the president of Italy on their behalf, say they are being accused of failing to do the impossible—predicting seismic events with certainty. The prosecution and many L'Aquila residents view the disaster quite differently, claiming that they do not fault the scientists qua scientists as much as they fault them in their role as government agents, whose job is to guard the public safety. In this, they purportedly failed. As a report on the trial in the journal *Nature* put the dilemma: "The legal fight in L'Aquila is viewed by some as a philosophical dispute between scientists, who believe that their role is pure hazard assessment, and the local prosecutors, who argue that Italian law obliges scientific advisers to evaluate the fragility of buildings and other factors in their assessment of risk."[1]

Setting questions of justice to the side for the moment, this dysfunctional drama—with the competing prophetic voices, the paradoxical expectations on science advisers, and the very real consequences of failing to meet them—should unfortunately be looking very familiar at this point in our genealogical survey of the scientific-prophetic ethos. Lest we are tempted to dismiss this story as a vagary of Italy's legal system, I would remind us of the federal and civil fraud investigations of Michael Mann, discussed in chapter 9, following his publication of the infamous "hockey stick" climate-change graph. Notwithstanding small successes in remediating their ethos, our science advisers clearly still face substantial ethical problems.

So now it is my turn to the play the oracle and prognosticate the future of the powerful but unstable ethos of the scientist-prophet. But first, as any good oracle would and should, I look at where we've been.

In defining the classical meme of prophetic ethos, I found it provided five consistent rhetorical motivations to the dramas in which it participated. The pivotal motivation was Ascertainment, the very purpose for which a polity consulted its prophet in the face of a political dilemma it could not resolve via normal political argumentation. The prophet used her privileged access to knowledge beyond the public ken to engage the polity in a dialogue. During this dialogue, the prophet led the polity to cast the political dilemma it faced it terms of its covenant values, the values that constituted it and distinguished it from other polities; this rhetorical fusion created a provisional political certainty that could drive political action.

Spoking off this central motivation, two other key dynamics structured the social context of classical prophecy: Authorization, the process by which the prophet was established by and outside the polity; and Confirmation, the politics of recognizing and legitimating the prophet's special status. Two further motivations drove the process of making prophetic texts: Divination, the techniques by which truth was made legible; and Prophecy, the framing and styling of an ongoing political dialogue about certainty.

We witnessed the first major crossing of scientific and prophetic ethos in the seventeenth century, first, as Francis Bacon argued for the institution of natural philosophy as a civic oracle, and later, as the Royal Society, ostentatiously taking up Bacon's mantle, used decidedly un-Baconian divinatory techniques to engage its peers, the court, and the clergy in a dialogue about the correct pursuit of certainty. This hybridization resulted in two important mutations to the classical prophetic meme: first, Ascertainment now aimed at natural rather than strictly theological truths; second, Divination was performed and validated communally.

While the Enlightenment appeared to push the Royal Society's prophetic program to its logical conclusion by establishing political regimes on "scientific" principles, it also threw a major wrench in the works of the progressive model for integrating scientific and prophetic ethos. Hume's Guillotine, or the is/ought stricture—the argument that you could not deduce from scientific observations of the world alone what should be done to manage that world, that values must be injected—laid the philosophical groundwork for an attempt to purify scientific ethos of its prophetic characteristics. In chapter 5, I showed how the is/ought stricture cashed out in rhetorical practice via stasis theory: (a) in a policy-making situation, the stasis system generates an anticipatory effect by which lower-stasis arguments are heard as higher-stasis arguments; (b) the is/ought boundary cuts through the stasis system at the pivotal stasis of cause and effect. These two rhetorical effects explain why science advisers' public arguments about the causes and effects of natural phenomena, such as climate change or stem cell research, are consistently heard as value judgments. Such transgressions of the is/ought ethical boundary are frequently punished, as we saw in chapters 6 and 9, particularly. My mapping of the scientific-prophetic discourse onto the stases also explains, I believe, previous observations of inevitable shifts up-stasis to evaluation when scientific findings are broadcast in the mass media: whenever we talk about the political applications of the privileged knowledge that scientists have brought back to us from the "wilderness" beyond our ken, we are engaging in prophecy.[2]

The philosophically simple ethical solution offered by the is/ought model—whereby scientists make claims within the proper bounds of their disciplinary ethos, and then politicians evaluate those claims and make policy from them—is rhetorically untenable.[3] Our increasingly technical societies have led us to rely increasingly on scientists for the instruction manual, so to speak. This pressure causes us to oscillate between the progressive and is/ought models in a cycle that Bruno Latour referred to as hybridization. Hybridization has rendered the ethos of the science adviser radically unstable—and science-policy making, frequently but not always, dysfunctional. What results is an ethical catch-22 for science advisers. Government agents call on these advisers to answer technical questions that have policy import; when the answers resonate with the current administration's policies, policy gets made relatively efficiently. But when outside critics want to challenge those policies, or when the advisors' recommendations run counter to administration policy, the is/ought stricture is conveniently invoked, and the advisors' prophetic ethos is used to undermine their disciplinary ethos.

We saw just such a case of ethical assassination in J. Robert Oppenheimer's federal service. Oppenheimer's bold performance of the progressive model of scientific-prophetic ethos, despite increasingly strong hints that this performance was making him enemies within the federal cult of Big Science, stands as this book's crowning tragedy. How different, then, Rachel Carson's ethical performance as the kairotic prophet, the voice "crying in the wilderness"—subject to the same pressures, her trajectory was nevertheless the opposite of Oppenheimer's: although initially his star was on the ascendant while she endured withering ethical critiques, the bright tail of her meteoric rise as the scientist of the people in the early 1960s crossed the hazy trail of his downfall. She ventriloquized the fears of marginalized mothers, retirees, and migrant farmworkers in the hallowed halls of Congress and the USDA, and the dialogue that she helped frame resulted in the formation of the EPA and still carries on.

The klieg lights that both Carson and Oppenheimer blinked into in the 1960s became footlights for the next act: science celebrities Stephen Hawking, Carl Sagan, Stephen Jay Gould, and a handful of other popularizers took the stage in the 1980s. They leveraged the powerful metaphor-maker of mass media—particularly film and TV—to circumvent traditional cultic bottlenecks of scientific communication and to take directly to lay polities their prophetic messages about the way that science and scientists should properly be employed. The mass media highlighted prophetic battles between science advisers and made tangible what we had sensed before—that what science produced as a public commodity was metaphor, and whoever controlled these metaphors increasingly controlled politics. But we also saw that controlling these metaphors was a Herculean labor.

That lesson continued to unfold in the 1990s and 2000s in debates about climate change. In our examination of the IPCC's ethical strategies in *Climate Change 2007,* we witnessed the report's authors deliberately disciplining uncertainty and deploying metaphoric GCM visualizations to defend their ethos. Savvy rhetoricians, these scientists appeared to have won a provisional stability for their ethos even though it was shaken by Climategate.

While the IPCC may have temporarily succeeded in remediating and stabilizing their scientific-prophetic ethos, we can predict based on the genealogy traced in this volume that they and oracles like them can continue to expect their pronouncements to be used as ammunition against their disciplinary ethos—no matter how much they protest that these pronouncements are "neutral with respect to policy."[4] The meme of scientific-prophetic ethos has been a long time in forming, and it will take more than a single performance—no matter how spectacular—to work a fundamental change in the schizophrenic way we deal with our science advisers.

Is such a memetic shift even possible? That is my purpose in this conclusion: to look at the proposals for remediating the ethos of the science adviser that have been offered to this point and to evaluate their feasibility based on the genealogy I have traced here.

Before we launch into the proposals, I should make clear my grounds for evaluation. I believe that many remedies proposed to this point evince a fundamental misunderstanding of the ethos of the science adviser: namely, they have framed it as *providing a technical solution to a political problem.* However, if we put the calling of science advisers in genealogical context, we realize that it is *to engage policymakers in a dialogue about certainty, a dialogue that casts the current dilemma in the light of covenant values.* Certainty, or the desire for it, forms the *grounds* of this calling but is not the *result* of it. What scientists-prophets do is to get politicians talking with them and with each other about the problem they face in terms of their core values. That's it. No deus ex machina, no magic bullet, no crystal ball.

Why has such a fundamental misunderstanding developed? Two reasons: first, we have bought the Enlightenment line about the disenchantment of the public sphere, so few scholars have thought to trace the ethos of the science adviser back to its premodern charismatic roots. Second, wishful thinking. In the face of the near-paralyzing technical complexity of some of the issues we face, we all—policymakers, parents, school board members, TV watchers alike—have elected to abdicate the hard decisions to the people with the technical knowledge. A lot of "science advising" at the governmental level amounts to "policy subcontracting." If we are tempted to feel righteous indignation against our elected officials, consider how many of us stopped eating eggs yolks when nutritionists reported they were loaded with cholesterol—and then started eating them again when they announced in the last decade that egg yolk cholesterol didn't appear to have any effect on human cholesterol? Did we understand the biochemistry behind either decision, or did we take both on blind trust? Are we feeling a little red in the face (and not just from high blood pressure)?

Subcontracting out policy decisions to technical advisers is just the progressive model in sheep's clothing, and it can work smoothly as it has in Germany with climate change legislation.[5] Politicians there have proven more willing to base their evaluations on values expressed by the scientific bodies advising them. Here in the United States, for better or for worse, we have insisted on a wider range of values being brought to the table—values including the precedence of industry and free trade, and the liberty of individuals (and corporations) to pursue their interests free from undue governmental intervention.

That these values do not derive from scientific polities does not mean that they have no bearing on scientific policy matters. One reason communication scholars remain unsure about how the public understands science is that they have tended to let scientists define "science" and scientific literacy for them.[6] By that measure, the public knows almost nothing and probably never will.[7] But carefully done surveys have shown that the American public in fact reads, thinks, and talks about science a great deal—through a "filter" of different values that makes this knowledge very difficult to compare to scientists' knowledge. These extrascientific values are formed on the basis of local social, cultural, economic, and physical experience.[8] While these local values cannot and should not dominate policy making on global issues like climate change, nonetheless, they index fine-grained information that scientific models miss. This information cannot be averaged or estimated but cannot be excluded from good policy making: some examples include the significance of salmon runs for Coastal American Indian religious practices or the circulation of banned pharmaceuticals in populations of illegal immigrants.[9]

It is in light of these two starting assumptions—that we have mistaken what our science advisers actually do for us, and that current models of the public understanding of science are working toward adequacy—that I will consider the various proposals for remediating the ethos of science advisers.

While these proposals are many, they fall into four basic categories: (1) proposals to strengthen the is/ought stricture on science-policy making, (2) proposals for a more progressive approach to science-policy making, (3) various educational proposals aimed at "narrowing" the knowledge gap between scientists and lay voters (this category comprises the educational arms of the first two categories; however, there are so many of these proposals they deserve their own discussion), and (4) proposals to reduce the influence of scientists in many kinds of science-policy making while increasing the influence of lay citizens and local communities.

Reinforcing the Is/Ought Divide

These proposals generally, but not always, come from conservative commentators. They also occasionally come from scientists who, irrespective of their political orientation, have gotten fed up with their bipolar treatment by politicians: fine, you stay on your side of the fence, and I'll stay on mine is the détente resorted to by these frustrated writers. Stephen Jay Gould's Non-Overlapping Magisteria (NOMA) was an example of this latter strategy, as is to a more moderate extent Roger Pielke Jr.'s "honest broker" proposal, which I discuss shortly in detail.

These NOMA-type proposals started almost as soon as Big Science policy making began in the United States. I referred to several in chapter 5. The most rhetorical of these was Marver Bernstein's, who, based on his examination of the (already) dysfunctional science-policy making going on, recommended that science advisers be restricted to policy implementation, not policy making, and that the technical questions put to them be carefully circumscribed beforehand by politicians.[10]

I think you can probably guess, having followed my argument to this point, how feasible I believe these sorts of proposals to be. Politicians by their own admission generally don't know enough of the technical subjects before them *to know what they don't know*,[11] so the precise framing of questions itself becomes impossible without science advisors; if that's the case, what are the chances of excluding scientists from policy making?

Roger Pielke Jr. offers a more moderate NOMA proposal tempered by his own extensive experience as a science adviser and as an observer and critic of the IPCC. He suggests the science adviser strengthen her disciplinary ethos by eliding the "I" or "we" from policy recommendations, instead presenting a range of policy options to policy makers without endorsing particular outcomes. Pielke calls this model the "honest broker" model and contrasts it with the "stealth issue advocacy" that can result when scientists present policy ("ought") as scientific fact ("is"); Pielke points to remote sensing of weapons of mass destruction in advance of the Iraq War as a particularly notorious case of a few technical experts in the Executive pushing policy with so-called "facts."[12] Pielke qualifies his recommendations by saying that all four models of science-advisory ethos—pure scientist, stealth issue advocate, issue advocate, and honest broker—need to be active for a truly healthy democratic science-policy making process; however, inside the halls of the Executive, Pielke appears to favor the honest broker as the best solution to the ethical catch-22 faced by science advisers.

The litmus test in these Executive situations for distinguishing the "stealth issue advocate" from the "honest broker of policy alternatives" is "the latter seeks to reduce the scope of available choice, while the former seeks to expand (or at least clarify) the scope of choice," a definition that sounds eminently reasonable but in fact raises an interesting problem.[13] Pielke's solution assumes two things: first, that the presentation of policy alternatives will constitute a convincing performance of disciplinary ethos, and second, that the science adviser can somehow anticipate all relevant policy alternatives. The first assumption may be logically sound, but as Pielke himself acknowledges elsewhere,[14] in practice a stunning amount of pressure is put on science advisers to reduce, not expand, policy options and to strengthen, not qualify, recommendations. So, these forces conspire against a successful performance of objectivity via the multiplying of options. As for the second assumption, Pielke again admits most policy situations feature "fundamental, irreducible uncertainty about the problem and policy options."[15] Given this situation, Pielke's "honest broker" will be forced to limit policy options using his or her best judgment. But by this act, she becomes vulnerable to the same criticisms leveled at "stealth issue advocates." We saw some of these criticisms begin to emerge, in fact, during Climategate when critics accused the IPCC of withholding or hiding alternative climate scenarios—even though they had already presented multiple such scenarios in the AR4.

To be clear, my argument here is not that Pielke's "honest broker" proposal, or others like it, would not on average work better than what we're doing now—just that it does not address the rhetorical fundamentals of the science advisors' fractured ethos.

Enforcing the Progressive Model

Proposals to enforce the progressive model of scientific-prophetic ethos *do* address the rhetorical drama created by the interference of the stases and the is/ought stricture—by doing away with the latter altogether. All these models begin with the assumption that scientists are better decision makers than the average voter, or that the covenant values of scientific polities (think of the Mertonian norms of disciplinary ethos) constitute the ideal covenant values of a pluralist democracy. These proposals accordingly seek to make scientific and political evaluative processes conform—either by making scientists into politicians or by making politicians into scientists.

Under the first model, scientists would be encouraged to leave their laboratories and classrooms for a stint in Congress; the aim is increasing the participation of technical professionals in Congress from the current 6% to something more nationally representative, 10% or so, or even to a higher fraction, as the organizers of the political action committee Ben Franklin's List hope.[16] The goal is for the values that scientists apply to policy decisions to infect Congress like a virus, leading to more scientific decision making across the board. However, it is hard to kick against the goads, as Bill Foster, a physicist working as a Representative for the state of Illinois in 2011, pointed out regarding the debt-ceiling debate: "The political logic is 'what I can get away with saying that people will believe'. . . . The scientific logic is 'what are the best estimates for the relevant numbers.' When the two collide . . . the political logic is overwhelming."[17]

Under the second model, scientists would mount an educational campaign to help politicians better understand technical policy issues. This is the charter of many scientific organizations, including the lobbying arms of the National Academy of Sciences and the American Association for the Advancement of Science. But both models assume that scientists and scientific values are closer to ideal than our current politicians and their covenant values are: this is the progressive core of these models. Whether you believe this argument or not is up to your personal politics, but I would submit that the question has been mooted by history. At the few junctures when significant science policy has been made on a progressive model—in the 1920s with the Wilson administration, in the 1940s with FDR, in the 1960s with Kennedy and, to some extent, Johnson, and in the 1990s with the Clinton administration—there have not been significantly more scientists or scientifically educated people in Congress. Therefore, these progressive spasms are likely due to the administrations in question being more willing to take the word of scientists on faith.[18] This record lends less support to progressive proposals for ethical integration than it does to conservative warnings about the development of an undemocratic "scientific-technological elite." I do not believe a technocracy is necessarily what Timothy Ferris and other advocates of progressive science-policy integration are shooting for.[19] And yet this is the inevitable result if the value and knowledge gap cannot be diminished between scientists and voters.

Educational Remedies

This discussion brings us to the key issue of science education, which some scholars propose as a partial solution to the current dysfunctional treatment of science advisers. To many, it is just common sense that if the voting public knew more about science, currently fractious decisions such as those about climate change or stem cell research would go more smoothly.

The oldest version of this proposal worked off what is now somewhat derisively called the "deficit model" by communications and education scholars. The model treated scientists and voters like two pitchers; all we needed to do was to fill the voters up to the scientists' levels—or at least to a level at which mutual understanding could occur. There were several problems with this model, one of which we have already discussed: the way the public "knows," evaluates, and uses science is very different from the way scientists do so, so definitions of scientific "literacy" remain contentious. Some education scholars believe "science literacy" itself has no empirical denotation, only a political one that perpetually points to a deficit in knowledge and thus to science-education funding as the solution.[20]

Most science-literacy advocates have now adopted more sophisticated proposals for remediating science-policy making via educational efforts.

Some of these proposals involve training scientists themselves to use mass media and to choose better metaphors to explain their work to nonspecialists: this is the goal of the Center for Communicating Science at Stony Brook University (which just added the actor Alan Alda to its faculty and features classes in improvisational comedy for scientists); related institutes are springing up almost on a daily basis to address a felt need for better communication of scientific knowledge.

Jon Miller's index of Civic Scientific Literacy shows that over the last 20 years the percentage of scientifically literate American adults has increased from 10% to 25%.[21] Although this looks like a distinct victory for educational efforts, formal or informal, keep in mind that scientific knowledge has been increasing more rapidly—by some estimates, *much* more rapidly, over the same period.[22] So, scientific ignorance may actually be increasing in spite of a concerted effort in this country since the Cold War began to improve scientific literacy. It now seems hopeless for the average American to keep up with even the basics of current scientific research, if that is indeed the goal.

The emphasis on metaphor in programs such as Stony Brook's promises to help bridge the knowledge gap with broad heuristics for understanding scientific advances. But we know enough now about the way that scientific metaphor works to see that even if voters succeed in understanding the metaphors, the resulting knowledge they acquire will be very different from the knowledge that trained scientists hold. In other words, the boundaries between scientific and nonscientific polities will remain solid, will perhaps even be reinforced. Americans will still essentially be taking scientific knowledge on trust.

But let us say that Americans in general somehow manage to acquire enough scientific competence to understand the science behind a proposed stem cell research policy using induced pluripotent stem cells—or that scientists manage to craft metaphors that successfully preserve the "immutable mobiles" they want to pass on concerning the policy—then what? The vast majority of science policy in this country is not voted on: it is made by appointed agents of federal agencies in consultation with science-advisory boards. True, we could write better-informed letters to our congressional representatives, who if so moved could call for policy review or could pass legislation to block the execution of our hypothetical stem cell policy; in much the same fashion, we could write letters to the president and hope for action from the Executive. In short, we could certainly get the attention of the people who make science policy in this country, but we are not they. This structural situation is not altered by our receiving more or better education or explanations about science. We still live, as far as science policy is concerned, in the technocracy we built in the first decades of the twentieth century.

As a last note on the educational programs, a frequent topos in discussions about education and the climate change debate is scientific uncertainty. The most common deployment of this topos holds that if audiences had a better appreciation of scientific uncertainty, they would better understand climate scientists' recommendations and would not demand the chimera of certainty; this version of the topos is particularly favored by climate scientists and is closely related to the deficit model of science education.[23] As such, it faces similar obstacles. As we have seen throughout this study, polities call on science advisers to manufacture *certainty* for them, not uncertainty. This framing of the drama of science advising makes nonscientist agents rhetorically tone deaf to arguments elaborating uncertainty. The IPCC discovered this firsthand when respondents to a reception survey of the TAR ignored explicit technical guidance about how to interpret statements of uncertainty such as "very likely" or "more likely than not."[24]

However, there is an alternate version of the scientific-uncertainty-education topos that concords better with the genealogy of the scientific-prophetic ethos I have developed in this study. This version recommends shifting the evaluation of uncertain issues like climate change from a scientific basis to a lay cultural basis. Lay citizens, the argument runs, are familiar with uncertainty in legal and economic situations and feel authorized to make decisions under it—such as rendering a verdict on a jury or deciding whether or not to buy a house.[25] To the extent that science-policy decisions are presented as types of decisions that are familiar and within public expertise, to that extent science-policy making will be more successful. This topos is elaborated by the fourth class of proposals, which recommends treating science-policy decisions as regular policy decisions—that is, as the province of citizens rather than of an elite extrapolitical class of experts.

More Seats at the Table—and Fewer of Them for Scientists

Since policy cannot be made without evaluation, and the values required must be supplied by the polity making the policy, some communication scholars have recommended shifting the focus in science-policy making off science advisers and back into the political arena. They justly point out that in this arena, scientists are citizens like everyone else, with their own cultural values and biases. As citizens, scientists command no special political expertise or authority, and they bear the same responsibilities for the outcome of science policy that their neighbors do. They should not have to bear more than their fair share of the decision-making responsibility; concomitantly, they should not wield more than their fair share of authority in the policy-making process.[26]

Policy under this model would be made entirely within the public ken using the common resources of democracy; scientists would thus perform expert but not prophetic ethos in this drama. Their knowledge and views would be weighted proportionally to their representation at the "table" of policy making, but no more.

Models like this have had limited success in addressing local environmental issues, where the polities affected can more easily identify themselves, articulate their covenant values, and agree on priorities.[27] But advocates of "seats at the table" reform, understandably, would like to see it happen in larger-scale issues such as climate change.[28] One factor in favor of these proposals is that climate change debates are now quite clearly stuck at a political rather than a technical stasis; that is, most scientists and most Americans agree that climate change is a real natural phenomenon; the "sticking points" now are all economic, social, and philosophical and thus fall in the jurisdiction of lay authority. However, a major obstacle to proposals to scale up "seats at the table" reform to the level of the climate change debate is that the table in question is a global one, and it is closeted in bits and pieces in the conference rooms of dozens if not hundreds of different nations' environmental agencies.[29] The UN might seem a promising forum in this regard, but it committed itself to the oracular model of science-policy making when it chartered the IPCC, and it is unlikely it will reform to embrace a more democratic model.

Conclusion: The Future of the Scientist-Prophet

In the end, I am not the Delphic oracle and cannot predict the future of science advising in America. What I do know after doing the research for this book is that the hybrid ethos of the scientist-prophet is held in its current form by a network of powerful rhetorical dynamics that are shaped themselves by longstanding social and political practices. This holding pattern would have to change significantly for the ethos of the science adviser to change in any significant way: "band-aid" solutions applied to the ethos of individual science advisers or advisory bodies will have no lasting effect. That being said, the proposals for remediating the currently unstable ethos of the American science adviser that have the best chance of success, in my judgment, are those that focus on increasing "seats at the table" in local policy decisions or those that depend on progressive administrations to instigate spasmodic periods of technocracy. Among all of the proposed solutions, only these two directly address the fundamental rhetorical dysfunction of the science adviser's ethos—by refusing to treat her as a prophet, as in the "seats at the table" model; or, by doing away with the is/ought stricture on her ethical performances, as in the progressive model.

If I can't make a prediction, can other scholars who have studied the long run of scientific ethos? In this area, I can think of no oracle better qualified than Steven Shapin, who has spent the better part of his career studying the history of scientific ethos. In his recent book, *The Scientific Life*, Shapin predicts we will return, in this era of cascading uncertainties, to a premodern system of placing trust in charismatic individuals. If such a memetic shift were to infect federal policy making, we would likely see the ethos of the science adviser revert to the progressive model, and we would be living in a technocracy. But Shapin's observations support another possible future in which science advising would be privatized, perhaps resembling special-interest lobbying. This scenario would most likely obtain in the event of a populist-conservative reorganization of the federal science agencies or a dramatic cutting of their research funding. Other scenarios, other kairoi, may open in response to as-yet-unforeseen shifts in political structures—only time will tell.

In the end, I may not know what the future holds, but I know this much: until we stop seeking certainty, we will consult prophets. And until some seemingly more-certain civic epistemology evolves,[30] we will seek our prophets among the sciences, and they will continue to engage us in a dialogue that goads us to recall our covenant values: we are the people who both love and fear science; we are the people who value industry and capital and yet want our natural environment to remain pristine; we are the people who prize solidarity and yet cannot bring ourselves to silence the voices speaking from the margin. And so, although our science advisers cannot tell the future or tell us what to do, they do—and will continue to—help us to know ourselves.

APPENDIX

KEY RECEPTION AND CONSTITUTION SOURCES

This appendix summarizes the type, amount, and source of primary evidence for claims made in this book about reception and constitution of scientific-prophetic ethos. Detailed reference information for these sources appears in the notes for each chapter.

Chapter 2: Commentary on Delphi and ancient Greek prophecy from contemporaries such as Cicero, Herodotus, and Plato. Also, Joseph Fontenrose's collection of Delphic pronouncements.

Chapter 3: Commentary on Bacon's ethos by contemporaries in letters, diaries, historical accounts. Primary texts by Bacon: *The Works of Francis Bacon* (14 vols.), ed. James Spedding, Robert Leslie Ellis, and Douglas Denon Heath. London: Longman, 1857–1874. Archive: Bacon Papers, Lambeth Palace Library, London.

Chapter 4: Correspondence among Society fellows (141 articles) reprinted in the first volume of the *Philosophical Transactions of the Royal Society* (1665–1666); commentary on the Society in contemporary plays, prose, and poetry; dialogues between some Society members and outside critics, particularly Joseph Glanvill/Henry Stubbe and Robert Boyle/Thomas Hobbes. Archives: Early English Books Online; Evelyn Family Papers, Hartlib Papers, and Oldenburg Papers at the British Library, London; the collections of the Royal Society Library, especially Birch's Council Minutes and the Boyle Papers.

Chapter 6: Reception by Gray Board and Atomic Energy Commission in hearings; mass media commentary on hearings and on Oppenheimer in general (roughly 600 clippings); constitution of scientists in mass media, particularly novels, films, and magazines. Archives: J. Robert Oppenheimer Papers, Library of Congress, Washington, DC; LexisNexis.Chapter 7: Constitution of Carson's ethos by PSAC and Ribicoff committee as recorded in transcripts and report drafts; reception of Carson and *Silent Spring* in contemporary mass media; secondary accounts of this reception in Lear, Murphy, and others. Archives: Rachel Carson Papers, Beinecke Library,

Yale University; Records of the Office of Science and Technology, National Archives, Washington, DC.

Chapter 8: Mass media reception of Oracles' media productions (roughly 30 reviews of each production discussed); constitution of Gould's ethos in ongoing creation/evolution debate as reflected in mass media and Arkansas trial transcript. Archive: LexisNexis.

Chapter 9: Dialogues between IPCC scientists and critics over AR4 and other reports, as reflected in roughly two dozen Internet blogs and web pages, three IPCC advisory documents, and one live presentation; debate over relationship of scientific uncertainty to scientific ethos since mid-1990s in about four dozen secondary sources; Google Timeline analytics for keywords "global warming denier."

NOTES

Notes on Abbreviations: After the first use of each, J. Robert Oppenheimer Papers, Library of Congress has been abbreviated "JRO Papers," and the Records of the Office of Science and Technology, National Archives has been abbreviated "OST."

PREFACE

1. Joseph Ben-David, *The Scientist's Role in Society: A Comparative Study* (Chicago: University of Chicago Press, 1984), 182.

CHAPTER 1

1. Michael Dowd, "In Time, Atheists May Come to Look like Prophets," *Commentary, Minnesota Public News*, Minnesota Public Radio, accessed September 20, 2011, http://minnesota.publicradio.org/display/web/2010/08/23/dowd/.
2. Blair Golson, "Sam Harris: The Truthdig Interview," *Truthdig* (blog), April 3, 2006, http://www.truthdig.com/report/page5/20060403_sam_harris_interview/.
3. Sam Harris and Annaka Harris, "Project Reason: A Non-Profit Dedicated to Reason," *Project Reason*, accessed September 20, 2011, http://www.project-reason.org/.
4. I derive my notion of polity from Aristotle's definition of it as a group of citizens who must use rhetoric to cooperate "for the common good" in *Politics*, trans. H. Rackham, Loeb Classical Library 264 (Cambridge, MA: Harvard University Press, 1926, 1999), 3.1279a. Although Aristotle elsewhere treated nondemocratic societies as polities, I choose to employ the democratic sense of polity throughout this book because it is the inability of a group of equals to decide what to do in a crisis that most frequently leads to consulting a prophet. Certainly, monarchs have employed prophets—as Nebuchadnezzar did Daniel, for instance—but these court prophets were personal consultants or instruments of hegemony, functioning quite differently from the interstate cults and itinerant prophets that Athens consulted. Since it is a commonplace that science and democracy

have developed together, the more strictly Athenian and democratic definitions of polity are more suitable to my project.

5. Thomas Goodnight, "The Personal, Technical, and Public Spheres of Argumentation: A Speculative Inquiry in the Art of Public Deliberation," in *Contemporary Rhetorical Theory: A Reader.*, ed. John Lucaites, Celeste Condit, and Sally Caudill (New York: The Guilford Press, 1999), 252.
6. "Ken" comes from an Old English word for "sight" or "view." I use it as shorthand for "the range of knowledge and action commonly available to a polity." My notion of public ken is perfectly compatible with Habermas's popular conception of the public sphere as far as political actions go: *Further Reflections on the Public Sphere* (Cambridge, MA: MIT Press, 1992). However, I add a dimension of civic epistemology that's not strictly part of Habermas's theory, and so my notion is more immediately inherited from Tom Goodnight's articulation of the private, public, and technological spheres in "The Personal, Technical, and Public Spheres of Argumentation: A Speculative Inquiry in the Art of Public Deliberation," 252. His "public sphere" is more equivalent to my "ken"; the "technological sphere" describes the special knowledge that scientists have that is beyond the public ken unless scientists make a concerted effort to translate it there across the sphere boundaries (which remain formidable).
7. Robert King Merton, *Sociology of Science: Theoretical and Empirical Investigations* (Chicago, IL: University of Chicago Press, 1973), 268. Merton's followers constituted a powerful school in the sociology of science, as I discuss in chapter 5. However, his theories came under criticism for their failure to account for variability in history and practice, and for their failure to distinguish "professed" ethical norms—those scientists express publicly—and "statistical" ethical norms—those they observe on a day-to-day basis in their work; see Barry Barnes and R. G. A. Dolby, "The Scientific Ethos: A Deviant Viewpoint," *European Journal of Sociology* 9 (1970): 3–25. With respect to this tradition, I note that I treat only professed norms of ethos in this book, that my construction of scientific-prophetic ethos crosses but does not follow Merton's closely, and that I believe as Barnes does that these norms change over time and space.
8. Michael J. Hyde, "Rhetorically, We Dwell," in *The Ethos of Rhetoric*, ed. Michael J. Hyde (Columbia: University of South Carolina Press, 2004), xiii–xxviii.
9. There are many systems for accounting for kairos. Classical sources treated kairos as an opportune moment, a chance to speak up and move political action forward; Onians derived his use of kairos from a cognate denoting the split second a weaver parted the warp threads to make a space for the shuttle to pass through—see Deborah Hawhee, *Bodily Arts: Rhetoric and Athletics in Ancient Greece* (Austin: University of Texas Press, 2004), 67. The rhetorical structures framing this opening were speaker, audience, and message. Twentieth-century scholars have expanded this "rhetorical triangle" to include *constraints* (persons,

objects, events, and relations) in the case of Lloyd Bitzer's "Rhetorical Situation." Lloyd Bitzer, "The Rhetorical Situation," *Philosophy and Rhetoric* 1 (1968): 1–14. Acknowledging nonhuman agents like computers, activity theory has added nodes such as tools, goals, and gatekeepers; see Donna J. Kain, "Constructing Genre: A Threefold Typology," *Technical Communication Quarterly* 14, no. 4 (2005): 375–409. Since I examine both ancient and modern kairoi, I need a system for accounting for kairos that is more sensitive to technology than the Classical triangle but less cumbersome than activity theory. The best balance for my purposes can be found in Kenneth Burke's pentad, which I will address shortly.

10. Leo Howe argues that all prediction is ideological in his introduction to *Predicting the Future*, ed. Leo Howe and Alan Wain (Cambridge, UK: Cambridge University Press, 1993), 6.
11. Lessl, Thomas M. *Rhetorical Darwinism: Religion, Evolution, and the Scientific Identity*. (Waco, TX: Baylor University Press, 2012).
12. I am indebted to Thomas Lessl for this articulation of science and scientists as a "public resource."
13. James Creese, "The Scientist in Politics" (paper presented at the Philadelphia Regional Technical Meeting of American Iron and Steel Institute, October 6, 1954); Harvey Brooks, "The Scientific Adviser," in *Scientists and National Policy-Making*, ed. Robert Gilpin and Christopher Wright (New York: Columbia University Press, 1964); Robert Gilpin, "Introduction: Natural Scientists in Policy-Making," in *Scientists and National Policy-Making*; Robert C. Wood, "Scientists and Politics: The Rise of an Apolitical Elite," in *Scientists and National Policy-Making;* Ralph Lapp, *The New Priesthood* (New York: Harper & Row, 1965); Joel Primack and Frank von Hippel, *Advice and Dissent: Scientists in the Political Arena* (New York: Basic Books, 1974); Judy Segal and Alan Richardson, "Scientific Ethos: Authority, Authorship, and Trust in the Sciences," *Configurations: A Journal of Literature, Science, and Technology* 11, no. 2 (2003): 137–44.
14. Some key science studies treatments of science-advising *ethos* include Sheila Jasanoff, *The Fifth Branch: Science Advisers as Policymakers* (Cambridge, MA: Harvard University Press, 1994); Naomi Oreskes, "Science and Public Policy: What's Proof Got to Do with It?" *Environmental Science & Policy* 7 (2004): 369–83; Roger A. Pielke, Jr., *The Honest Broker: Making Sense of Science in Policy and Politics* (New York: Cambridge University Press, 2007); Steven Shapin, *The Scientific Life*.
15. Charles Austin Beard and William Beard, *The American Leviathan: The Republic in the Machine Age* (New York: The Macmillan Company, 1930); Timothy Ferris, *The Science of Liberty: Democracy, Reason, and the Laws of Nature* (New York: Harper Perennial, 2010); Charles Percy Snow, *Science and Government* (Cambridge, MA: Harvard University Press, 1961).

16. Talcott Parsons, ed., *Max Weber: The Theory of Social and Economic Organization*, trans. A. M. Henderson and Talcott Parsons (New York: Oxford University Press, 1947), 386–90.
17. See Thomas M. Lessl, "Gnostic Scientism and the Prohibition of Questions," *Rhetoric & Public Affairs* 5, no. 1 (2002): 133–57. Robert K. Merton traced the monastic behaviors of the scientists he studied back to the ascetic Puritan ethos in "The Normative Structure of Science," in *The Sociology of Science: Theoretical and Empirical Investigations*, ed. Norman W. Storer (Chicago: University of Chicago Press, 1973). Michel Foucault argued that before the Enlightenment truth could only be attained via a combination of *askēsis*, ascetic self-destruction, and *ēros*, a divine gift. Ancient philosophers and prophets sacrificed themselves to guide people in these spiritual-ethical pursuits. *Hermeneutics of the Subject: Lectures at the Collège De France, 1981–82* (New York: Picador Press, 2005).
18. Max Weber, "Science as a Vocation," in *Max Weber: Essays in Sociology*, ed. H. H. Gerth and C. Wright Mills (1919; Oxford, UK: Oxford University Press, 1946), 155.
19. Bruno Latour, *We Have Never Been Modern*, trans. Catherine Porter (Cambridge, MA: Harvard University Press, 1993).
20. Thomas M. Lessl, "Heresy, Orthodoxy, and the Politics of Science," *Quarterly Journal of Speech* 74, no. 1 (1988): 18–34; Lessl, "Gnostic Scientism and the Prohibition of Questions."
21. John W. McReynolds, "Note on the Compleat Expert," *Bulletin of the Atomic Scientists* 7, no. 1 (1951): 15–16.
22. Kenneth Burke calls this mode of rhetorical inquiry "ancestral" or "familial" in *Grammar of Motives* (Berkeley: University of California Press, 1969), 28, 402.
23. Michel Foucault, "Nietzsche, Genealogy, History," in *Language, Counter-Memory, Practice: Selected Essays and Interviews*, ed. Donald F. Bouchard (Ithaca, NY: Cornell University Press, 1980).
24. Ibid., 164.
25. Susan Blackmore, "Imitation and the Definition of a Meme," *Journal of Memetics—Evolutionary Models of Information Transmission* 2 (1998), http://jom-emit.cfpm.org/1998/vol2/ blackmore_s.html.
26. See chapter 11 in Richard Dawkins, *The Selfish Gene*, 30th anniversary ed. (New York: Oxford University Press, 2006).
27. Zi-Ran He, "Linguistic Memes and Their Rhetoric Effects," *Foreign Language Research* 1 (2008), http://en.cnki.com.cn/Article_en/CJFDTOTAL-OUTL200801012.htm; Davi Johnson, "Mapping the Meme: A Geographical Approach to Materialist Rhetorical Criticism," *Communication and Critical/Cultural Studies* 4, no. 1 (2007): 27–50.
28. David K. Dirlam, "Competing Memes Analysis," *Journal of Memetics—Evolutionary Models of Information Transmission* 7 (2003), http://cfpm.org/jom-emit/2003/vol7/dirlam_dk.html; James Gleick, "What Defines a Meme?"

Smithsonian, May 2011, 3, http://www.smithsonianmag.com/arts-culture/What-Defines-a-Meme.html?c=y&page=1.

29. I go into this distinction in chapter 5.
30. Burke, *Grammar of Motives*, 38.
31. Kenneth Burke, *Language as Symbolic Action* (Berkeley: University of California Press, 1966), 6.
32. Burke, *Grammar of Motives*, xx–xxi.
33. Leah Ceccarelli, *Shaping Science with Rhetoric: The Cases of Dobzhansky, Schrödinger, and Wilson* (Chicago, IL: University of Chicago Press, 2001), 7.
34. Beard and Beard, *The American Leviathan;* Vannevar Bush, *Science the Endless Frontier*, ed. Office of Scientific Research and Development (Washington, DC: Government Printing Office, 1945); Ferris, *The Science of Liberty*; Merton, *Sociology of Science*; Parsons, *Max Weber: The Theory of Social and Economic Organization*.
35. My justification and motivation for this dominantly Western focus is that classical Greco-Roman and Judeo-Christian prophecy were the biggest influences on performances of prophetic ethos in seventeenth-century London, which I take as the first major crossing of prophetic and scientific ethos in history (although I recognize that the Italian Renaissance supported significant attempts to hybridize prophecy and natural philosophy—more on this in chapter 3). In terms of non-Western prophetic traditions, several sociologists and anthropologists have worked and are working in this area, including Holger Jebens, *Cargo, Cult, and Culture Critique* (Honolulu: University of Hawai'i Press, 2004); and Thomas Overholt, *The Channels of Prophecy: The Social Dynamics of Prophetic Activity* (Minneapolis, MN: Fortress Press, 1989). Some have even investigated the connections between prophecy and science—to wit, studies of astronomical observations inscribed on Chinese oracle bones from the Shang dynasty in D. N. Keightley, *Sources of Shang History: The Oracle-Bone Inscriptions of Bronze Age China* (Berkeley: University of California Press, 1985). However, to account for all global traditions of prophecy in the space of this book would be impossible, and at any rate, the founding members of the flagship scientific societies in Europe had access only to the most fragmentary information about these traditions. These fragments they sometimes recorded in their commonplace books as curiosities, as did John Evelyn regarding Turkish magicians in "Locorum Comunium Tomus 1mus," in Evelyn Papers, MS Add 78328 (London: British Library). However Evelyn and his colleagues could not investigate these tidbits systematically enough to integrate them into their practices in any meaningful way.
36. John L. Pauley II has offered a brief rhetorical definition, but his focus is a particular kairos in African American politics; I work from the considerable anthropological and sociological literature on historical and worldwide prophetic practices to provide a more broadly applicable definition of prophetic ethos.

CHAPTER 2

1. Plato, *Apology* (Cambridge, MA: Harvard University Press, 1966), 113–15.
2. Pausanias, *Description of Greece* (Cambridge, MA: Harvard University Press, 1918), http://www.perseus.tufts.edu/.
3. Sarah Iles Johnston, *Ancient Greek Divination*, Blackwell Ancient Religions (Chichester, UK: Wiley-Blackwell, 2008), 35.
4. Page DuBois, *Sowing the Body: Psychoanalysis and Ancient Representations of Women*. (Chicago, IL: University of Chicago Press, 1988), 97.
5. Joseph Blenkinsopp, *A History of Prophecy in Israel* (Philadelphia, PA: Westminster Press, 1983), 36; Robert R. Wilson, *Prophecy and Society in Ancient Israel* (Philadelphia, PA: Fortress Press, 1980), 23.
6. Herodotus, *Herodotus, with an English Translation by A. D. Godley* (Cambridge, MA: Harvard University Press, 1920), http://www.perseus.tufts.edu/.
7. Aristotle, *Metaphysics* (Cambridge, MA: Harvard University Press, 1989), 9.1050b, http:// www.perseus.tufts.edu/. Aristotle centered his theory of being on the actual and potential states of all objects. He coined a term to denote what he means by actuality, *entelecheia*, which blended the words *energeia* ("operation," derived from *ergos*) and *tēlos* ("perfection," or ultimate end). Thus, *entelecheia* acknowledged the current form of things but also accounted for the energy expended in maintaining that form. The word Aristotle uses for potentiality is *dunamis*, from which we get our "dynamic/s."
8. Marcus Tullius Cicero, "De Divinatione," Loeb Classical Library 154 (Cambridge, MA: Harvard University Press, 1923), 2.21–25, 75.
9. Margaret D. Zulick, "The Agon of Jeremiah: On the Dialogic Invention of Prophetic Ethos," *Quarterly Journal of Speech* 78 (1992): 125–48.
10. Sarah Iles Johnston, *Religions of the Ancient World: A Guide* (Cambridge, MA: Belknap Press of Harvard University Press, 2004), 248.
11. Blenkinsopp, *A History of Prophecy in Israel*, 59.
12. Jebens, *Cargo, Cult, and Culture Critique*.
13. Bruno Snell, *The Discovery of the Mind* (New York: Harper and Brothers, 1960).
14. Isocrates, "Nicocles," trans. George Norlin, Loeb Classical Library 209 (Cambridge, MA: Harvard University Press, 1928), 8.
15. Isocrates, "Antidosis," trans. George Norlin, Loeb Classical Library 229 (Cambridge, MA: Harvard University Press, 1929, 1996), 295–97; Henri Irenée Marrou, *A History of Education in Antiquity*, trans. George Lamb, 3rd ed., Wisconsin Studies in Classics (Madison: University of Wisconsin, 1982), 88.
16. Joseph Fontenrose, *The Delphic Oracle: Its Responses and Operations with a Catalogue of Responses* (Berkeley: University of California Press, 1978), 48.
17. In this section, I develop the way I am using the term "Other" along contemporary lines. But readers familiar with Hegel's conception of the Other, or Edward Said's development of it in postcolonial theory, will be comfortable with the

way it develops here. For readers unfamiliar with the term, the Other is a psychological, social, racial, or political entity we recognize as different than and potentially threatening to the Self.

18. Still today in Greece, prophecy is the office of women, who, in communities such as Inner Mani, are responsible for reading death omens and performing ritual laments: C. Nadia Seremetakis, *The Last Word: Women, Death, and Divination in Inner Mani* (Chicago, IL: University of Chicago Press, 1991), 17.
19. Euripides, "Captive Melanippe," in *Euripides: Selected Fragmentary Plays*, ed. C. Collard, M. J. Cropp, and K. H. Lee (Warminster, UK: Aris & Phillips, 1995), 494.12–17.
20. Marie Mauxion, "La Transmission De La Parole Oraculaire," *Langages* 85 (1987): 9–14.
21. Fontenrose, *The Delphic Oracle: Its Responses and Operations with a Catalogue of Responses*, 145; Mauxion, "La Transmission De La Parole Oraculaire," 316.
22. Karlyn Kohrs Campbell, "Consciousness-Raising: Linking Theory, Criticism, and Practice," *Rhetoric Society Quarterly* 32, no. 1 (2002): 52–3.
23. Zulick, "The Agon of Jeremiah: On the Dialogic Invention of Prophetic Ethos."
24. Plato, *Phaedrus* (Cambridge, MA: Harvard University Press, 1925), 244C, http://www.perseus.tufts.edu/.
25. Plato, *Symposium*, trans. W. R. M. Lamb, Loeb Classical Library 166 (Cambridge, MA: Harvard University Press, 1925), 201D.
26. Strabo, *Geography of Strabo*, trans. Horace Leonard Jones, Loeb Classical Library 196 (Cambridge, MA: Harvard University Press, 1924), 9.7.
27. Blenkinsopp, *A History of Prophecy in Israel*; Wilson, *Prophecy and Society in Ancient Israel*; Johnston, *Ancient Greek Divination*; Burke O. Long, "Prophetic Authority as Social Reality," in *Canon and Authority: Essays in Old Testament Religion and Theology*, ed. George W. Coats and Burke O. Long (Philadelphia, PA: Fortress Press, 1977), 1–20; Zulick, "The Agon of Jeremiah: On the Dialogic Invention of Prophetic Ethos."
28. Blenkinsopp, *History of Prophecy in Israel*, 57.
29. Ibid., 41; Johnston, *Ancient Greek Divination*, 117.
30. Blenkinsopp, *History of Prophecy in Israel*, 64; Lester L. Grabbe, *Priests, Prophets, Diviners, Sages: A Socio-Historical Study of Religious Specialists in Ancient Israel* (Valley Forge, PA: Trinity Press International, 1995), 115. Women prophets are particularly well-attested in the ancient Hebrew texts, which relate the powers and prophecies of Miriam, Deborah, and Huldah alongside their male counterparts with no distinction other than the gendering of their title: nebîʾāh as opposed to the male nābî.
31. Plutarch, *The Obsolescence of Oracles*, trans. Frank Cole Babbitt, Loeb Classical Library 306 (New York: G.P. Putnam, 1927), 438B–C.
32. *Heraclitus*, ed. Phillip Ellis Wheelwright (Oxford, UK: Oxford University Press, 1959), 70–71.

33. Whether or not the Pythia entered a divine frenzy (*enthousiasmos*) during consultations has been a matter of debate since Plato claimed so in the *Phaedrus* (244a–b). Plutarch, Strabo, Herodotus, Lucan (see Lucan, *The Pharsalia of Lucan*, [London: Longmans, Green, and Co., 1896], 190–232, http://omacl.org/Pharsalia/book5.html) and Cicero took sides on the issue in classical times, and it has had remarkable staying power. Scholars in the 1960s and 1970s argued strenuously for a rational, coherent Pythia. Then in the early twenty-first century, geo-archaelogical work found evidence of the hallucinatory gas ethylene in young faults crossing directly beneath the *adyton*. Jelle Z. de Boer, John R. Hale, and Jeffrey Chanton, "New Evidence for the Geological Origins of the Ancient Delphic Oracle (Greece)," *Geology* 29, no. 8 (2001): 707–10. For a rhetorician, what is interesting is not trying to settle historical issues like this but rather examining the strategies that contemporary writers used in arguing about them. Those who espoused the Pythia's mania used it as evidence that the human cosmos was bounded by chaos and thus was neither omniscient nor omnipresent.
34. These details are taken from contemporary accounts including the following (relevant passages cited throughout this chapter): Herodotus's *Histories*; Strabo's *Geography*; Plutarch's *Moralia* (*The E at Delphi, The Obsolence of Oracles, and the Oracles at Delphi*). The best summaries of the contemporary record are in Bowden, *Classical Athens and the Delphic Oracle*; Johnston, *Ancient Greek Divination*.
35. Thucydides, *The Peloponnesian War* (New York: E. P. Dutton, 1910), http://www.perseus.tufts.edu/.
36. Overholt, *The Channels of Prophecy: The Social Dynamics of Prophetic Activity*, 70–76.
37. Wilson, *Prophecy and Society in Ancient Israel*, 47.
38. Johnston, *Ancient Greek Divination*, 112.
39. Ibid., 63.
40. Wilson, *Prophecy and Society in Ancient Israel*, 49.
41. Blenkinsopp, *A History of Prophecy in Israel*, 30; Brian Peckham, *History and Prophecy: The Development of Late Judean Literary Traditions* (New York: Doubleday, 1993), 134; David L. Petersen, *The Prophetic Literature: An Introduction* (Louisville, KY: Westminster/John Knox Press, 2002), 32.
42. Timothy Polk, "The Prophetic Persona: Jeremiah and the Language of Self, *Journal for the Study of the Old Testament*, supplement Series 32 (Sheffield, UK: JSOT Press, 1984), 173; Overholt, *The Channels of Prophecy: The Social Dynamics of Prophetic Activity*, 88.
43. Wilson, *Prophecy and Society in Ancient Israel*, 56.
44. An example of how this type of interaction sounded: Isyllos the poet petitioned the Pythia around 350 B.C. concerning a paean he wished to compose in honor

of Apollo. Isyllos asked, "Is it better [*lōion kai ameinōn*] that he inscribe the paean?" And the Pythia responded, "It is better [*lōion kai ameinōn*] for both present and future that he inscribe the paean." Fontenrose, *The Delphic Oracle: Its Responses and Operations with a Catalogue of Responses*, 252.

45. Johnston, *Ancient Greek Divination*, 54. Also, see Cicero, *On Divination* (1.49).
46. *Anagignoskein* was ambiguous between reading and "recognizing," a conflation which bears on our discussion of the politics of Confirmation in prophecy.
47. For examples of this kind of advice, see Aristotle, *Aristotle on Rhetoric: A Theory of Civic Discourse*, ed. George Kennedy (New York: Oxford University press, 1991), 2.22; Gorgias of Leontini, *Encomium of Helen* (Bristol, UK: Bristol Classical Press, 1982).
48. Overholt, *The Channels of Prophecy: The Social Dynamics of Prophetic Activity*, 117.
49. *Hymn to Hermes* (Cambridge, MA: Harvard University Press, 1914), 521–73, http://www.perseus.tufts.edu/.
50. Johnston, *Ancient Greek Divination*, 99–100, 131.
51. Demosthenes, *Demosthenes with an English Translation*, trans. A. T. Murray (Cambridge, MA: Harvard University Press, 1939), 43.66–67, http://www.perseus.tufts.edu/.
52. Martin Heidegger, *On the Way to Language*, trans. Peter D. Hertz (New York: Harper & Row, 1971), 29.
53. Hans Georg Gadamer, *On Truth and Method* (London, UK: Continuum, 2004), 296.
54. Grabbe, *Priests, Prophets, Diviners, Sages: A Socio-Historical Study of Religious Specialists in Ancient Israel*, 64; Overholt, *The Channels of Prophecy: The Social Dynamics of Prophetic Activity*, 118; Wilson, *Prophecy and Society in Ancient Israel*, 160.
55. Euripides, *Cyclops, Alcestis, Medea*, trans. David Kovacs, Loeb Classical Library 12 (Cambridge, MA: Harvard University Press, 1994), 679–81.
56. *The E at Delphi*, 386E; *The Sign of Socrates*, 579B–D.
57. Heraclitus's theory of language held in general that antinomies could not hold up in action. His fragments include several vivid analogies along these lines. One contended that sea water was both poison and life, but this paradox was not apparent until it was consumed (by people and by fish, respectively) (101). Another example: the concept of health was clear to people only when they became sick (99). One of Heraclitus's most famous demonstrations of the fundamental instability of antinomies in the face of action involved a "barley-drink" (*kykeon*), which had to be agitated to count as *kykeon*; in stasis it was merely separate layers of honey-water, grated cheese, and barley (50).
58. Polk, *The Prophetic Persona: Jeremiah and the Language of Self*, 173–74; Zulick, "The Agon of Jeremiah: On the Dialogic Invention of Prophetic Ethos."

59. Charles Bazerman, *Shaping Written Knowledge: The Genre and Activity of the Experimental Article in Science* (Madison: University of Wisconsin Press, 1988); Carolyn R. Miller, "Rhetorical Community: The Cultural Basis of Genre," in *Genre and the New Rhetoric*, ed. Aviva Freedman and Peter Medway (London: Taylor & Francis, 1995), 57–66. This definition of genre may sound like my definition of a meme; in fact, prophetic genres are one part of the larger meme of prophetic ethos.
60. Peckham, *History and Prophecy*, 11, 144.
61. Polk, *The Prophetic Persona: Jeremiah and the Language of Self*, 173–74.
62. Bowden, *Classical Athens and the Delphic Oracle*, 4.

CHAPTER 3

1. Lisa Jardine credits this experiment to Bacon in "Monuments and Microscopes: Scientific Thinking on a Grand Scale in the Early Royal Society," *Notes & Records of the Royal Society of London* 55 (2001): 302n. Given that he cribbed much of the content of *Inquiry Respecting the Magnet* from William Gilbert's treatise *De Magnete* (although not this experiment in particular), it is just as likely that Bacon recorded it from another source.
2. Ibid., 290.
3. William Rawley, ed. *Manes Verulamiani: (Shades of Verulam)* (New York: Bacon Society of America, 1927), 7. For sixteenth- and seventeenth-century English texts in this chapter and the next—both manuscript and printed—I have regularized print styling and orthography (e.g., transcribed words printed or written as *Iustice* to "justice") but have otherwise let contemporary spellings stand, as they aren't that difficult for the modern reader to make out.
4. In 1594, Bacon, in yet another attempt to secure a permanent position in Elizabeth's administration, wrote her favorite, Robert Devereux, the Earl of Essex: "Therefore my L[ordship] I was determined and am determined if her majestie reject me . . . I will by goddes assistance with this digrace of my fortune and yet with the comfort of the good opinion of so many honorable and worthy persons retire my self with a couple of men to Cambridge and theare spend my life in my studies and contemplations without looking backe. I humbly pray your L[ordship] so pardon me for troubling you with my melancholy. For the matter it self, I comend it to your love." Francis Bacon, Letter from Francis Bacon to the Earl of Essex, 1594, folder 94, MS 649, Bacon MSS, Lambeth Palace Library. On this same theme, Bacon wrote Thomas Bodley bemoaning his greatest error in life, "That knowing my Self, by inward Calling, to be fitter, to hold a Book, than to play a part, I have led my life in Civil Causes; for which I was not very fit by nature, and more unfit, by the preoccupation of my Minde." William Rawley, *Resuscitatio, or, Bringing into Publick Light Severall Pieces, of the Works, Civli, Historical, Philosophical, & Theological, Hitherto Sleeping; of the Right*

Honorable Francis Bacon Baron of Verulam, Viscout Saint Alban. According to the Corrected Coppies. Together, with His Lordships Life (London: Sarah Griffin, for William Lee, 1657), 34.

5. This is the judgment of all his major biographers, including James Spedding in the nineteenth century and, most recently, Brian Vickers and Markku Peltonen in their biographical notes to the *Major Works* and the *Cambridge Companion*, respectively. None protests that Bacon was entirely innocent; rather, all argue that gift giving was such an extensive practice at James I's court that Bacon's being singled out for it likely had other political motives. The King was under a great deal of pressure from Parliament to reform his government in 1621, and Bacon as Lord Chancellor was a high-visibility scapegoat.
6. *The Works of Ben Jonson* (Boston: Phillips, Sampson, & Co., 1853), 834.
7. Thomas Sprat, *The History of the Royal-Society of London, for the Improving of Natural Knowledge.* (London: T. R. for J. Martyn at the Bell without Temple-bar, 1667), v.
8. Thomas M. Lessl, *Rhetorical Darwinism: Religion, Evolution, and the Scientific Identity* (Waco, TX: Baylor University Press, 2012), 56.
9. In her recent study of scientific activity in Elizabethan society, Deborah Harkness rightly points out that historians have mistakenly deemed Bacon the "prophet" of early modern science when in fact it was already well underway; see *The Jewel House: Elizabethan London and the Scientific Revolution*, (New Haven, CT: Yale University Press, 2007), 8. She does admit that he was far more successful than his contemporaries in promoting experimental science as an activity worthy of state support (252). This is the facet of Bacon's rhetoric that I am interested in—his prophetic construction of the natural philosopher and the reception and re-performance of that ethos by the founders of the Royal Society.
10. See Markku Peltonen, introduction to *The Cambridge Companion to Bacon*, ed. Markku Peltonen (Cambridge, UK, and New York: Cambridge University Press, 1996) for a cogent summary of this historiography as well as a sample of the best recent approaches to Bacon's philosophy and rhetoric. For a classification of scholarly approaches to Bacon's scientific rhetoric, see also James P. Zappen, "Francis Bacon and the Historiography of Scientific Rhetoric," *Rhetoric Review* 8, no. 1 (1989): 74–88.
11. This is essentially Lessl's justification for focusing on Bacon as a nexus in the development of the "priestly" ethos of the scientist. I don't disagree: I don't need this justification. The Royal Society publicly and repeatedly pointed to Bacon as its prophetic forebear, and that tradition justifies my decision to look at how Bacon hybridized prophetic and scientific ethos. For the term "civic epistemology," see Sheila Jasanoff, "Cosmopolitan Knowledge: Climate Science and Global Civic Epistemology," in *The Oxford Handbook of Climate Change and Society*, ed. John S. Dryzek, Richard B. Norgaard and David Schlosberg, 127–43 (Oxford, UK: Oxford University Press, 2011).

12. Quotes from *Works* (4.26, 33) and Benjamin Farrington, *Philosophy of Francis Bacon* (Chicago, IL: University of Chicago Press, 1964), 94–98, 133. The way Bacon used the term "oracle" is particularly interesting. By my count, he used it 50 times (throughout Spedding's authoritative *Works*). Most often, he used the word to refer to the queen or king's judgment. For instance, in his "Speech on Taking His Seat in Chancery," he assured the peerage that in case of divided opinion, "I should have recourse to the oracle of the King's own judgment before I should pronounce" (*Letters,* VI, 187). Such language would not have been considered blasphemous at this time because the Divine Right of Kings was still assumed. Perhaps because of this provenance, Bacon never used the word "oracle" to refer to himself. But he frequently called himself a prophet or *interpreter* of oracles. As an epitome of this type of usage, in the trial against Essex in 1601, Bacon bemoaned "how true a prophet" he was for having foreseen the disaster that would ensue from Essex's Irish campaign (10.146). Bacon is not here referring to biblical prophecy; he is constructing a sort of political prophecy that combines encyclopedic knowledge of history and law with astute observation of present circumstances at Court. To sum up, Bacon used the topoi of oracles and prophecy in his political writings to (a) construct himself as a loyal servant of the Crown and (b) cast the search for truth as a corporate political enterprise. These features of political prophecy served to discipline the independence of the ethos of the natural magician, as we will see shortly.
13. Quoted in Francis Bacon, Lisa Jardine, and Michael Silverthorne, *The New Organon* (Cambridge, UK: Cambridge University Press, 2000), xxviii.
14. Natural was not about hexes and familiars: writers in Bacon's time would have called that kind of magic "spiritual" or "demonic." Richard Sergeantson, "Natural Knowledge in the New Atlantis," in *Francis Bacon's New Atlantis: New Interdisciplinary Essays,* ed. Bronwen Price (Manchester, UK: Manchester University Press, 2002), 88.
15. Sir Alan Cook, "Essay Review: Seeds of the Scientific Revolution," *Notes & Records of the Royal Society of London* 51 (1997), 159–60; Stephen Gaukroger, *Francis Bacon and the Transformation of Early-Modern Philosophy* (Cambridge: Cambridge University Press, 2001), 175–95; John S. Mebane, *Renaissance Magic & the Return of the Golden Age,* (Lincoln: University of Nebraska Press, 1989); Paolo Rossi, *Francis Bacon: From Magic to Science,* trans. Sacha Rabinovitch (London 1968); Sergeantson, "Natural Knowledge in the New Atlantis"; Ryan J. Stark, *Rhetoric, Science, and Magic in Seventeenth-Century England* (Washington, DC: Catholic University of America Press, 2009); Frances Yates, *Giordano Bruno and the Hermetic Tradition* (London: Routledge, 1964); Perez Zagorin, *Francis Bacon* (Princeton, NJ: Princeton University Press, 1998), 28–29.
16. Yates, *Giordano Bruno and the Hermetic Tradition,* 52.
17. Mebane, *Renaissance Magic and the Return of the Golden Age,* 22.

18. "Behold now, standing before you, the man who has pierced the air and penetrated the sky, wended his way amongst the stars and overpassed the margins of the world, who has broken down those imaginary divisions between spheres—the first, the eighth, the ninth, the tenth, or what you will—which are described in the false mathematics of blind and popular philosophy." Quoted ibid., 87.
19. Ibid., 83.
20. See Stark, *Rhetoric, Science, and Magic in Seventeenth-Century England*, 10–18.
21. Gaukroger, *Francis Bacon and the Transformation of Early-Modern Philosophy*, 175–95; Mebane, *Renaissance Magic & the Return of the Golden Age*, 7; Rossi, *Francis Bacon: From Magic to Science;* Sergeantson, "Natural Knowledge in the New Atlantis"; Zagorin, *Francis Bacon*, 40–43.
22. John C. Briggs, "'*The Very Idea!*' Francis Bacon and E. O. Wilson on the Rehabilitation of Eidos," in *Francis Bacon and the Refiguring of Early Modern Thought: Essays to Commemorate the Advancement of Learning (1605–2005)*, ed. Julie Robin Solomon and Catherine Gimelli Martin (Hants, UK: Ashgate, 2005), 94–97; Mebane, *Renaissance Magic & the Return of the Golden Age*, 22; Rossi, *Francis Bacon: From Magic to Science*, 16.
23. Rossi, *Francis Bacon: From Magic to Science*, 12.
24. Daniel Edward Seward, "Civic Voice in Elizabethan Parliamentary Oratory: The Rhetoric and Composition of Speeches Delivered at Westminster in 1566" (PhD diss., University of Texas, 2008), 291.
25. Cicero, Aristotle, and Plato were widely read in the English Renaissance and after, as were Plutarch, Xenophon, Herodotus, Strabo, Lucan, and the other Classical historians I consulted in chapter 2. See Thomas Whitfield Baldwin, *William Shakespere's Small Latine and Lesse Greeke* (Champaign, IL: University of Illinois Press, 1944), 380–408, http://durer.press.illinois.edu/baldwin/index.html; Peter Mack, *Elizabethan Rhetoric: Theory and Practice*, Ideas in Context (Cambridge, UK: Cambridge University Press, 2002), 13, 66. Mythology was a favored genre for personal reading and pedagogy in Bacon's time as well; see Barbara Carman Garner, "Francis Bacon, Natalis Comes, and the Mythological Tradition," *Journal of the Warburg and Courtauld Institutes* 33 (1970): 264–91.
26. I follow Barry Brummett in distinguishing millennialist from millenarian rhetorical strategies as follows: a millennialist rhetoric supposes that the Millennium has arrived (or is in the process of unfolding); it therefore works within the establishment, calling for the individual to align herself with the newly revealed order of history, which is progressing ever upward toward a final, vaguely defined moment of apocalypse. By contrast, a millenarian rhetoric reads the current historical crises as presaging imminent apocalypse, which will cleanse history and usher in the (more vaguely defined) Millennium. Both modes were active in the era under discussion; the millenarian strand was particularly strong in epochs when Catholics, such as Mary, returned to the throne.

But in toto, Brummett identifies Bacon's and the later Puritan reforms of philosophy as millennialist, operating within the established order to bring about the progressive improvement of common life; see *Contemporary Apocalyptic Rhetoric,* Praeger Series in Political Communication, ed. Robert E. Denton Jr. (New York: Prager, 1991), particularly chapter 3.

27. Seward, "Civic Voice in Elizabethan Parliamentary Oratory: The Rhetoric and Composition of Speeches Delivered at Westminster in 1566," 291.
28. John N. King, "Robert Crowley's Editions of 'Piers Plowman': A Tudor Apocalypse," *Modern Philology* 73, no. 4 (1976): 344.
29. David Norbrook, *Poetry and Politics in the English Renaissance* (London: Routledge & Kegan Paul, 1984), 35–53. Crowley published a number of tracts and "gospelling" poems on the theme of Edward's prophetic destiny. A few examples: in the *Waie of Wealth,* Crowley promises Protestants that Edward will free them from oppression by Papists: "And then he that caused King Cirus to send the Iewes home to Ierusalem againe shall also stire up our yong king Edward to restore thee to thy liberty againe." R. Crowley and J. M. Cowper, *The Select Works of Robert Crowley: Printer, Archdeacon of Hereford (1559–1657), Vicar of St. Lawrence, Jewry, &C. &C.* (London: Pub. for the Early English Text Society, by N. Trübner & Co., 1872), 137. And, in his commentary on his edition of *Piers Plowman,* which recasts it as an apocalyptic fable, Crowley predicts of Edward's rule, "And kind love shall come yet, and conscience togither And make of lawe a labourer, such love shall aryse And such a peace among the people, & a perfite truth." King, "Robert Crowley's Editions of "Piers Plowman:" A Tudor Apocalypse," 350. John Bale's commentary on Revelation makes similar arguments, and at its conclusion he exhorts his Protestant readers to pray daily for "that most worthy minister of God, Kyng Henry the eight...which hath so sore wounded the beast, that he may before his departure, or Prynce Edwarde after him, throw all his supersticyons into the bottomlesse lake agayne (from whence they have commen) to the comforte of his people." John Bale, *The Image of Both Churches after the Most Wonderfull and Heauenly Reuelation of Sainct Iohn the Euangelist, Contayning a Very Fruitfull Exposition or Paraphrase Vpon the Same. Wherin It Is Conferred Vvith the Other Scriptures, and Most Auctorised Histories. Compyled by Iohn Bale an Exyle Also in Thys Lyfe, for the Faithfull Testimony of Iesu.* (London: Thomas East, 1550, 1570), 144r.
30. William Perkins, *The Arte of Prophecying, or, a Treatise Concerning the Sacred and Onely True Manner and Methode of Preaching First Written in Latine by Master William Perkins; and Now Faithfully Translated into English (for That It Containeth Many Worthie Things Fit for the Knowledge of Men of All Degrees) by Thomas Tuke* (London: Felix Kyngston for E. E. and are to be sold in Pauls Church-yard at the signe of the Swan., 1607), 148, Early English Books Online.
31. Francis Bacon, *The Advancement of Learning,* in *The Major Works,* ed. Brian Vickers (New York: Oxford University Press, 2002), 128.

32. Thomas M. Lessl, "Francis Bacon, the Advancement of Learning, and the Biblical Origins of the Scientific Ethos," *Journal of Communication and Religion* 15 (1992): 93.
33. Bacon, *The Advancement of Learning*, 142.
34. Ibid., 124.
35. Seward, "Civic Voice in Elizabethan Parliamentary Oratory: The Rhetoric and Composition of Speeches Delivered at Westminster in 1566," 295.
36. Mebane, *Renaissance Magic & the Return of the Golden Age*, 83; Charles Webster, *The Great Instauration: Science, Medicine and Reform, 1626–1660* (London, 1975), 16.
37. Farrington, *Philosophy of Francis Bacon*, 65.
38. Rossi, *Francis Bacon: From Magic to Science*, 14.
39. Antonio Pérez-Ramos, "Bacon's Forms and the Maker's Knowledge Tradition," in *The Cambridge Companion to Bacon*, ed. Markku Peltonen (Cambridge, UK: Cambridge University Press, 1996), 103.
40. Farrington, *Philosophy of Francis Bacon*, 55.
41. Pérez-Ramos, "Bacon's Forms and the Maker's Knowledge Tradition," 101. In his best-selling collection of contemporary glosses on ancient myths, the *Wisdom of the Ancients*, Bacon puts a "maker's knowledge" spin on a favored symbol of the Hermeticists—"The Sphinx": "Moreover the riddles of the Sphinx have always a twofold condition attached to them; distraction and laceration of mind, if you fail to solve them; if you succeed, a kingdom. For he who understands his subject is master of his end; and every workman is king over his work... For the command over things natural,—over bodies, medicines, mechanical powers, and infinite other of the kind—is the one proper and ultimate end of true natural philosophy" (6.161).
42. The *New Atlantis* was published posthumously along with the *Sylva Sylvarum* and went through 17 editions by the end of the seventeenth century. Lessl, *Rhetorical Darwinism: Religion, Evolution, and the Scientific Identity*, 77.
43. Francis Bacon, *The New Atlantis*, in *The Major Works*, ed. Brian Vickers (New York: Oxford University Press, 2002), 461.
44. Ibid., 471, 480.
45. Ibid., 478.
46. Ibid., 487.
47. Ibid., 480–81.
48. Ibid., 488.
49. Steven Shapin, *A Social History of Truth: Civility and Science in Seventeenth-Century England* (Chicago, IL, and London, UK: University of Chicago Press, 1994), 46.
50. As further support for my argument here, I will point out that a recent compilation of essays returned to *New Atlantis* and found evidence that the fable

amplified marginalized voices. Bronwen Price, ed. *Francis Bacon's New Atlantis: New Interdisciplinary Essays* (Manchester, UK: Manchester University Press, 2002). See in particular Bronwen Price's introductory essay and Kate Aughterson's chapter on gender.

51. Bacon, "The New Atlantis," 480–87.
52. Lessl, *Rhetorical Darwinism: Religion, Evolution, and the Scientific Identity*, 77.
53. "Francis Bacon, the *Advancement of Learning*, and the Biblical Origins of the Scientific Ethos," 97. Bacon makes the "two books" argument in nearly every one of his treatises on natural philosophy: *Masculine Birth of Time*, in Farrington, *Philosophy of Francis Bacon*, 60, 65–66; *Valerius Terminus* (3.29); the *Advancement of Learning* (126, 153); *Wisdom of the Ancients* (6.98); *Thoughts and Conclusions on the Interpretation of Nature* (Farrington 78–79); and the *Great Instauration* (4.20).
54. In the *Advancement*, Bacon subjugated natural philosophy to theology on the grounds that first causes resided in the mind of God. Nevertheless, in an interesting passage in the *Advancement*, Bacon argued that human, natural, and divine philosophy all aimed at what he called *Philosophia Prima* (189–91). He did not elaborate *Philosophia Prima* in any detail, but it seemed to indicate the fundamental method for apprehending truth of any kind and involved operations such as noticing difference and similarity. So, although natural philosophy was subordinate to theology, *Philosophia Prima* could not be correctly constituted without it. See Sachiko Kusukawa, "Bacon's Classification of Knowledge," in *The Cambridge Companion to Bacon*, ed. Markku Peltonen (Cambridge, UK: Cambridge University Press, 1996), 54–55.
55. Simon Wortham, "Censorship and the Institution of Knowledge in Bacon's *New Atlantis*," in *Francis Bacon's New Atlantis: New Interdisciplinary Essays*, ed. Bronwen Price (Manchester, UK: Manchester University Press, 2002), 186.
56. Cornelius Agrippa, *Of the Vanitie and uncertaintie of Artes and Sciences*, 1569, Early English Books Online.
57. Lessl, *Rhetorical Darwinism: Religion, Evolution, and the Scientific Identity*, 54.
58. Farrington, *Philosophy of Francis Bacon*, 60.
59. Ibid., 67–68.
60. Ibid., 69.
61. Francis Bacon, *Sylva Sylvarum or a Natural History in Ten Centuries Written by The Right Honorable Francis Lord Verulam Viscount St. Alban. Published after the Author's Death by William Rawley Dr. Of Divinity, &C.* (London: W. Lee, 1627), A2.
62. It is worth noting that Bacon classifies divination in Book II of the *Advancement* as a sub-type of human learning under investigations of the mind. First, he divides divination as did Cicero into natural and artificial divination, and then, "For artificial divination, the several kinds thereof are distributed amongst particular knowledges. The Astronomer hath his predictions, as of conjunctions, aspects, eclipses, and the like. The Physician hath his predictions, of death, of recovery, of the accidents and issues of diseases" (216). Bacon does not

recommend the methods of these workers due to their lack of systematicity, but he approves of their objects: he includes natural divinations pertaining to weather, earthquakes, etc. in his ideal experiments at the end of *New Atlantis.*

63. See Gaukroger, *Francis Bacon and the Transformation of Early-Modern Philosophy*, 102–04, 128–29; Harkness, *The Jewel House: Elizabethan London and the Scientific Revolution*, 253; Michel Malherbe, "Bacon's Method of Science," in *The Cambridge Companion to Bacon*, ed. Markku Peltonen (Cambridge, UK, and New York: Cambridge University Press, 1996), 81; Pérez-Ramos, *"Bacon's Forms and the Maker's Knowledge Tradition,"* 110; Paolo Rossi, "Bacon's Idea of Science," in *The Cambridge Companion to Bacon*, ed. Markku Peltonen (Cambridge, UK, and New York: Cambridge University Press, 1996), 31–38.
64. "For even as in the business of life a man's disposition and the secret workings of his mind and affections are better discovered when he is in trouble than at other times, so likewise the secrets of nature reveal themselves more readily under the vexations of art than when they go their own way" (4.95).
65. Using heat as an exemplary nature, first a worker would record all results indicating the presence of heat—fire, cow dung, wet cotton, etc.; then, for each instance of presence, the worker would note any further experiments on it that led to the absence of heat—pouring water on fire, for example; finally, the worker would list conditions in which heat was found to increase or decrease.
66. Under his discussion of rhetorical method ("judgment," or what Ramus would have called *dispositio*) in the *Advancement*, Bacon thus distinguished fable and aphorism: "The pretence" of fable "is to remove the vulgar capacities from being admitted to the secrets of knowledges, and to reserve them to selected auditors, or wits of such sharpness as can pierce the veil" (234); aphorism, on the other hand,

    a. trieth the writer, whether he be superficial or solid: for Aphorisms, except they should be ridiculous, cannot be made but of the pith and heart of sciences; for discourse of illustration is cut off; recitals of examples are cut off; discourse of connexion and order is cut off; descriptions of practice are cut off; so there remaineth nothing to fill the Aphorisms but some good quantity of observation: and therefore no man can suffice, nor in reason will attempt to write Aphorisms, but he that is sound and grounded. (234–5)

    By this description, aphorisms are terse and direct while fables are allusive and indirect. But later in his discussion of political history, Bacon lumped the methods together with parables, maxims, axioms, and "sentences politic" as the ancients' preferred strategy for communicating "any observation that they thought was good for life" (270).
67. Bacon, *The Advancement of Learning*, 235.
68. In fact, the *New Atlantis* and the *New Organon* reworked in parabolic form some rather bald recommendations for the authorization of science that Bacon had made in his unpublished tract *Thoughts and Reflections on the Interpretation of*

Nature. Benjamin Farrington suspects *Thoughts* remained unpublished because draft reviewers (Thomas Bodley and Bishop Lancelot Andrewes) took exception to its directness (See Farrington, *Philosophy of Francis Bacon*, 46.). Committed to these ideas, Bacon may have taken his own advice about using "similitudes" to soften up his audience.

69. James Zappen and Kate Aughterson have reviewed this literature and found three opposing positions: (1) Bacon indeed advocated plain style, and the Royal Society adopted his recommendations; thus, to this day, scientists still eschew ornate prose. This conclusion was largely reached by Bacon enthusiasts and scientific apologists in the nineteenth and early twentieth centuries. (2) Bacon advocated plain style but did not use it, and neither did his followers. This was the position of revisionist historians in the 1960s through 1980s. (3) Bacon in fact was not advocating plain style, just a judicious and controlled use of figurative language, but he has been misread by many scientists and historians since due to their own biases concerning the social influence of science. This has been the position of the most recent work, which Zappen argues is aimed at opening the discourse of science to more participants and interpretations. Again, I am not interested in settling this debate but rather in looking at what Bacon argued about plain style and trying to explain why he might have argued that way given the political climate and his investment in prophetic ethos. See Kate Aughterson, "Redefining the Plain Style: Francis Bacon, Linguistic Extension, and Semantic Change in *The Advancement of Learning*," *Studies in Philology* 97, no. 1 (2000): 96–143; Zappen, "Francis Bacon and the Historiography of Scientific Rhetoric."
70. He makes further comments on plain style in the *De Dignitate et Augmentis Scientiarum*, the Latin translation and expansion of Book II of the *Advancement* that Bacon published in 1623 as part of his *Great Instauration* project.
71. *Perkins, The Arte of Prophecying, or, a Treatise Concerning the Sacred and Onely True Manner and Methode of Preaching First Written in Latine by Master William Perkins; and Now Faithfully Translated into English (for That It Containeth Many Worthie Things Fit for the Knowledge of Men of All Degrees) by Thomas Tuke*, 22, 97, Early English Books Online.

CHAPTER 4

1. There were earlier informal and/or sporadic scientific organizations in Europe, including the short-lived Accademia dei Lincei in Italy. And, several were officially recognized shortly after the Royal Society, including the French Académie Royale des Sciences (1666) and the German Leopoldina (1687). But the Royal Society still has precedence as the earliest official scientific society still in operation.

2. Michael Hunter, *Establishing the New Science: The Experience of the Early Royal Society* (Woodbridge, UK: Boydell Press, 1989), 199–229; Richard Kroll, Richard Ashcraft, and Perez Zagorin, eds., *Philosophy, Science, and Religion in England 1640–1700* (New York: Cambridge University Press, 1992), 10–11.
3. Peter Dear, "Totius in Verba: Rhetoric and Authority in the Early Royal Society," *Isis* 76 (1985): 147; Malherbe, "Bacon's Method of Science," 75.
4. Bacon, *Works*, 14:544.
5. Prophetic ethos does not necessarily address antinomies every time it is performed. This is not a requirement of Burke's model but rather a side effect of his having built and tested it on classical drama. While the pentad always reveals dilemmas or crises, they will not always be binary. That being said, in the seventeenth-century *kairoi* I consider, the Agents were all educated on a classical paradigm (see Mack, *Elizabethan Rhetoric: Theory and Practice*, 9), and so it was quite natural for them to frame the crises they faced in antinomies, much to Bacon's annoyance; cf. Bacon, *The Advancement of Learning*, 141. So, that particular dynamic of ancient prophetic *ethos* still obtains for the group I study in the next two chapters.
6. Keith Wrightson, *Earthly Necessities: Economic Lives in Early Modern Britain* (New Haven, CT: Yale University Press, 2000), 230.
7. Marcelo Gleiser, *The Prophet and the Astronomer: Apocalyptic Science and the End of the World* (New York: Norton, 2001), 30–31; James Kinsley, "The 'Three Glorious Victories' in Annus Mirabilis," *The Review of English Studies VII*, no. 25 (1956); Tim Thornton, *Prophecy, Politics and the People in Early Modern England* (Boydell, 2006), 99–102; David Wilson, "The Myth of Mother Shipton: Prophet-Making and Profit-Taking," in *Shell Games: Studies in Scams, Frauds, and Deceits (1300–1650)*, ed. Margaret Reeves, Richard Raiswell, and Mark Crane (Toronto, CA: Centre for Reformation and Renaissance Studies, 2004), 309–24.
8. Henry G. Van Leeuwen, The Problem of Certainty in English Thought 1630–1690, *International Archives of the History of Ideas* (The Hague: Martinus Nijhoff, 1970), chapter 3.
9. John Wilkins, *Of the Principles and Duties of Natural Religion Two Books/ by the Right Reverend Father in God, John, Late Lord Bishop of Chester* (London, 1675), 1, Early English Books Online.
10. Joseph Glanvill, *Scepsis Scientifica, or Confess'd Ignorance, the way to science in an essay of The vanity of dogmatizing, and confident opinion: with a reply to the exceptions of the learned Thomas Albius* (London, 1665), 4, Early English Books Online.
11. Joseph Glanvill, *Plus Ultra: or the Progress and Advancement of Knowledge Since the Days of Aristotle. In an Account of some of the most Remarkable Late Improvements of Practical, Useful Learning: to Encourage Philosophical Endeavours. Occasioned By a conference with one of the notional Way* (London, 1665), 148, Early English

Books Online. We recall "enthusiasm" from chapter 3 as the heretical practice of espousing individual, direct revelation in defiance of Church authority.

12. Thomas Birch, *Journal Books of the Royal Society: Volume 1, 1660–1826*, MS GB 0117 JBC, *Royal Society*, March 18, 1663.
13. Ibid., July 22, 1663.
14. Steven Shapin and Simon Schaffer, *Leviathan and the Air-Pump: Hobbes, Boyle, and the Experimental Life* (Princeton, NJ: Princeton University Press, 1985), 25.
15. Volume I of the Society's Journal Books records these early ongoing dialogues and interactions centering on the experiments that were performed on (roughly) a weekly basis at Trinity House.
16. Shapin and Schaffer, *Leviathan and the Air-Pump*, 315.
17. Ibid., 340.
18. Cf. John Beale's "tryals" of spring water by boiling and tasting it (*PTRS* 1:359–360) or Thomas Henshaw's tests of May-dew's purported powers to cure complexion (1:33–36). In the rest of this chapter, I will follow this convention for citing articles from the *PTRS* with the volume number (1) followed by the page numbers.
19. Dava Sobel, *Longitude: The True Story of a Lone Genius Who Solved the Greatest Scientific Problem of His Time* (New York: Penguin Books, 1996).
20. Shapin and Schaffer, *Leviathan and the Air-Pump*, 21.
21. Jan Comenius, *Copy Letters in Scribal Hand, Comenius to Morian and Hartlib, English Translation of Latin Originals*, 7 March/17 February 1641, folder 84, MS 7, pp. 1A-4B. Hartlib Papers, British Library.
22. Gaukroger, *Francis Bacon and the Transformation of Early-Modern Philosophy*, 131.
23. R. F. Jones, *Ancients and Moderns: A Study of the Rise of the Scientific Movement in Seventeenth-Century England* (Berkeley: University of California Press, 1962), 115–17; Webster, *The Great Instauration: Science, Medicine and Reform, 1626–1660*, xv–xvi.
24. John Milton, *Areopagitica; a Speech of Mr. John Milton for the Liberty of Unlicens'd Printing, to the Parlament of England* (London,1644), 31, Early English Books Online. Of the Interregnum, Ryan Stark has argued "It is difficult to overstate the extent to which eschatology influenced English theology and rhetoric" (*Rhetoric, Science, and Magic in Seventeenth-Century England*, 77). This revived millenarianism re-energized arguments linking apocalypse and operative natural magic, two popular expressions of which can be found in William Lilly's *Englands Propheticall Merline* (1644) and John Gadbury's *Nativity of the Late King Charls* (1659). Such an argument was quite naturally picked up by advocates of philosophical reform, chief among whom was John Webster; his *Academiarum examen* (1653) was enormously influential on the Hartlib circle; see R. H. Syfret, "Some Early Reactions to the Royal Society," *Notes and Records of the Royal Society of London* 7, no. 2 (April 1, 1950): 223.

25. John Beale, *Memo on a College, folder 1f, MS 31,* pp. 77A–80B. Hartlib Papers, British Library; John Evelyn, *Letter to Honorable H. Robert Boyle, 1657, folder Add. 4229,* p. 52. Evelyn Papers, British Library; Samuel Hartlib, *Letter Hartlib to Robert Boyle, 8 May 1654,* MS BOYLE/5, pp. 83–89. Hartlib Papers, British Library.
26. John Dury, *A Brief Discourse Concerning the Accomplishments of Our Reformation Shewing That by an Office of Addresse in Spirituall and Temporall Matters the Glory of God and the Happines of This Nation May Be Highly Advanced, 1646, folder 10, MS 17,* pp. 1A–20B. Hartlib Papers, British Library.
27. Hunter, *Establishing the New Science: The Experience of the Early Royal Society,* 17.
28. Stephen Wren, *Parentalia: Or, Memoirs of the Family of the Wrens; Viz. Of Mathew Bishop of Ely, Christopher Dean of Windsor, &C. But Chiefly of Sir Christopher Wren, Late Surveyor-General of the Royal Buildings, President of the Royal Society, &C. &C. In Which Is Contained, Besides His Works, a Great Number of Original Papers and Records; on Religion, Politicks, Anatomy, Matehmaticks, Architecutre, Antiquities; and Most Branches of Polite Literature. Compiled by His Son Christopher;* (London: T. Osborn, in Grays-Inn; and R. Dodsley, in Pall-Mall, 1750), 196.
29. Ibid., 197–98.
30. The "Orphean Charm" passage alludes to an idea Bacon explicated in the *Wisdom of the Ancients* under "Orpheus, or Philosophy": "The singing of Orpheus is of two kinds: one to propitiate the infernal powers, the other to draw the wild beasts and the woods. The former may be best understood as referring to natural philosophy; the latter to philosophy moral and civil" (6.111). We're in a position to note that both charms echo Strabo's explanation of how the Delphic oracle led to the founding of democracy by requiring tribes to coordinate their support of the shrine.
31. The founders of the Royal Society made so many Baconian references and allusions that I can only give a sample here. I have already covered the arguments for Salomon's college specifically, though I will add that Joseph Glanvill heralded the *New Atlantis* as a "prophetick scheam" in *Scepsis Scientifica,* C2r. Principally, we see arguments that the founders were continuing Bacon's general *Advancement of Learning*: for instance, this beautiful passage from one of John Beale's letters, "Others can only wonder at things rare & distant. I am broken into wonder at things obvious & under my feet. I cannot find, why the best grasse is soe purely greene; some starvy grasse of a russet Sea-greene. Lord Bacon hath allmost taught me to know, why the snowe is soe white." Extract in Scribal Hand, Beale To ?Hartlib, 1658, MS RS/BL09. Hartlib Papers, British Library. See also Robert Boyle, *The Excellency of Theology, Compar'd with Natural Philosophy (as Both Are Objects of Men's Study.) Discours'd of in a Letter to a Friend* (London: Printed by T. N. for Henry Herringman, at the Anchor in the Lower Walk of the New Exchange, 1674), 6; John Evelyn, *Locorum Communarium Tomus Iidus,* MS Add 78329. Evelyn Papers, British Library; *Letter to Dr. William Wotton, 1696,*

folder Add. 4229, pp. 58–59b. Evelyn Papers, British Library; Joseph Glanvill, *Plus Ultra*, 8, 86–88, 115; A. R. Hall and M. B. Hall, eds., *Correspondence of Henry Oldenburg*, vol. 1 (Madison, WI: University of Wisconsin Press, 1966), 87; Robert Hooke, *Micrographia: Or Some Physciological Descriptiosn of Minute Bodies Made by Magnifying Glasses. With Observations and Inquiries Thereupon* (London: Printed by John Martyn, and Jame Allestry, Printers to the Royal Society, 1665), 7; Sprat, *The History of the Royal-Society of London, for the Improving of Natural Knowledge*, 35.

32. Wrightson, *Earthly Necessities: Economic Lives in Early Modern Britain*, 202.
33. Ibid., 210, 303.
34. Bacon's secretary had to protest on his behalf in the introduction to the *Sylva Sylvarum*, "And as for the baseness of many of the experiments; As long as they be God's works, they are honourable enough" in *Sylva Sylvarum or a Natural History in Ten Centuries Written by The Right Honorable Francis Lord Verulam Viscount St. Alban. Published after the Author's Death by William Rawley Dr. Of Divinity, &C.*, A2r. Boyle expended several pages proving experiments did not degrade the gentleman in *Some Considerations Touching the Usefulnesse of Experimental Naturall Philosophy, Propos'd in a Familiar Discourse to a Friend, by Way of Invitation to the Study of It* (Oxford: Printed by Henry Hall Printer to the University, for Richard Davis, 1664), 10–30; he did it again in *Christian Virtuoso: Shewing That by Being Addicted to Experimental Philosophy, a Man Is Rather Assisted Than Indisposed to Be a Good Christian*, 40–65.
35. Charles Sylvester, "Leisure, Science, and Religion in 17th-Century England," *Leisure Sciences* 16, no. 1 (1994): 1–16.
36. Daniel Seward, "Financing God's Monarchy: Religious Inflections on Classical Deliberative Rhetoric" (paper presented at the International Society for the History of Rhetoric, Montreal, July 23, 2009).
37. I am indebted to Dan Seward for suggesting that this was the primary antinomy the founders confronted in their Authorization efforts.
38. Shapin, *A Social History of Truth*, 396.
39. Harkness, *The Jewel House: Elizabethan London and the Scientific Revolution*, 211–53.
40. As a reminder from chapter 2, following sociohistorical work on religious intermediaries, I do not treat "prophet" and "priest" as distinct *ethē* but rather as different emphasis within the ethos of the religious intermediary—prophecy and cultic administration, respectively. I use "prophet" as the cover term in this book because I am primarily concerned with religious intermediaries known as much or more for their words as their other actions. Such a label covers the divinatory and prophetic functions exercised by priests as well.
41. Boyle, *Some Considerations*, 19–22, 48.

42. R. H. Syfret, "Some Early Reactions to the Royal Society," 223. The examples from primary sources are too numerous and too long to quote here, but see Boyle, *Some Considerations; Walter Charleton, The Darkness of Atheism Dispelled by the Light of Nature. A Physico-Theological Treatise* (London: J.R. for William Lee, 1652); John Evelyn, *Letter to Robert Boyle, September 1659,* folder Add. 4263, pp. 69–72b. Evelyn Papers, British Library; the Preface of Glanvill, *Plus Ultra;* the Preface of Hooke, *Micrographia*; and the "Dedicatory Epistle" of Wilkins, *An Essay Towards a Real Character, and a Philosophical Language.*
43. Boyle, *Some Considerations*, 54–55.
44. Gilbert Burnet, "A Sermon Preached at the Funeral of the Hon. Robert Boyle," in *Lives, Characters, and a Sermon Preached at the Funeral of the Hon. Robert Boyle* (Dublin: Printed for A. Watson, 7 Capel-Street, 1824), 247; Shapin, *A Social History of Truth*, 178.
45. Shapin, *A Social History of Truth*, 42–125.
46. Ibid., 392–407.
47. Jardine, "Monuments and Microscopes: Scientific Thinking on a Grand Scale in the Early Royal Society," 302.
48. Samuel Pepys, *Diary of Samuel Pepys* (Project Gutenberg, 2004), 21 January 1667, http://www.gutenberg.org/dirs/4/1/7/4176/4176.txt.
49. Hooke prepared an important and widely discussed proposal for reforming the Society's working methods in 1673; see Hunter, *Establishing the New Science: The Experience of the Early Royal Society*, 216–26. Close to 1 in 10 articles in the first volume of the *PTRS* is attributed to or quotes "the ingenious Mr. Hook." Finally, Henry Oldenburg, the *PTRS* editor, and the Society were embroiled for some time in Hooke's ultimately unsuccessful attempts to guard his patent on the balance-spring watch: Hooke, *Proposalls for the Good of the Royal Society*, MS CP/20/50, Royal Society; see also his *Lectiones Cutlerianae, or a Collection of Lectures: Physcial, Mechanical, Geographical, & Astronomical. Made before the Royal Society on Several Occasions at Gresham Colledge. To Which Are Added Divers Miscellaneous Discourses* (London: Printed for John Martyn Printer to the Royal Society, at the Bell in S. Pauls Church-yard, 1679). This record confirms Hooke's pivotal role in the Society's day-to-day operations.
50. Hooke, *Micrographia*, 1.
51. Harkness, *The Jewel House: Elizabethan London and the Scientific Revolution*, 253.
52. Shapin, *A Social History of Truth*, 371.
53. David F. Noble, *A World without Women: The Christian Clerical Culture of Western Science* (New York: Alfred Knopf, 1992), 163.
54. *"Mv Lady Ranelagh's Choice Receipts, as also some of Captain Wilks, who valued them above gold,"* Sloane MSS 1367, British Library.
55. Margaret Cavendish, *Observations upon Experimental Philosophy* (1666), http://burton.byu.edu/courses/291/cavendish.htm.

56. Pepys, *Diary of Samuel Pepys*, 30 May 1667.
57. Shapin uses this term to refer to the Society's invisible assistants Shapin, *A Social History of Truth*, 372.
58. Ibid., section 3.
59. Richard Holmes, "The Royal Society's Lost Women Scientists," *The Observer: News Review*, November 20, 2010, 12. In addition to Hooke's famous vivisection, some of the experiments recorded in the first volume include cutting off strips of animals' skin, pouring salt on slugs, injecting viper poison into cats and dogs, pricking kittens with poisoned daggers, and bleeding dogs until their hearts stopped beating. There is a tragicomic moment recorded in the Council minutes wherein "[t]he dog, that had a piece of his skin cut off at the former meeting, being inquired after, and the operator answering, that he had run away, it was ordered, that another should be provided against the next meeting for the like experiment." Birch, *Journal Books of the Royal Society*: Volume 1, 22 June 1663.
60. See the dedication from Margaret Cavendish, *The Description of a New World, Called the Blazing World* (London: A. Maxwell, 1668), http://digital.library.upenn.edu/women/newcastle/blazing/blazing.html.
61. Ibid., para. 27.
62. *The World's Olio* (London: Printed for J. Martin and J Allestrye at the Bell in St. Paul's Churchyard, 1655), 7, Early English Books Online.
63. Ibid., 5.
64. Richard Nate, "'Plain and Vulgarly Express'd': Margaret Cavendish and the Discourse of the New Science," *Rhetorica: A Journal of the History of Rhetoric* 19, no. 4 (Autumn 2001): 404.
65. Holmes, "The Royal Society's Lost Women Scientists."
66. The Society's records themselves corroborate Holmes's argument about the persistent and important, though faint, presence of women in the constitution of experimental ethos. Women were employed as divinatory instruments to sense light, sound, smell, etc. in experiments since they were thought to be more sensitive to the fringes of perception: Jan Golinski mentions several such experiments involving the perception of the glowing of phosphorent material in "A Noble Spectacle: Phosphorus and the Public Cultures of Science in the Early Royal Society," *Isis* 80, no. 1 (1989): 11–39. And Boyle himself reported an instance in which he was given an idea for an experiment by a "very fair" young woman who pointed out that she could attract her hair to her skin by rubbing it; in another experiment, a young woman in his household is called in to scent the air while the philosophers open a vacuum in which a rose has been preserved. Michael Hunter and C. Littleton, "The Work-Diaries of Robert Boyle: A Newly Discovered Source and Its Internet Publication," *Notes & Records of the Royal Society of London* 55 (2001): 378.

67. I find it more than coincidental that Bacon and the early founders persistently figured Nature as a woman: a wife for "generation, fruit, and comfort" (Bacon, *The Advancement of Learning,* 148); a queen who "to be commanded must be obeyed" (Spedding, Ellis, and Heath, eds., *Works of Francis Bacon,* 47); a "pregnant" guardian of secrets (John Beale, *Letter, 13 July 1667,* folder 4294, pp. 34–35b) who must be "vexed" (Bacon, *Works,* 4.95), "penetrated" (Boyle, *The Excellency of Theology, Compar'd with Natural Philosophy (as Both Are Objects of Men's Study.) Discours'd of in a Letter to a Friend,* 193), and otherwise raped of them. She was the chaotic Other to the Society's totalizing Self; making her intelligible required *askēsis,* self-sacrifice, as I will discuss in the next section.
68. Noble, *A World without Women,* 208.
69. See Joseph Glanvill, *Saducismus triumphatus: or, Full and plain Evidence Concerning Witches and Apparitions* (London, 1681), Early English Books Online; and, unpublished manuscripts by Boyle, including a letter on witchcraft to Glanvill quoted in Michael C. W. Hunter, *Robert Boyle, 1627–91: Scrupulosity and Science* (Boydell Press, 2000), 102.
70. Robert Boyle, *New Experiments Physico-Mechanicall, Touching the Spring of the Air, and Its Effects, (Made, for the Most Part, in a New Pneumatical Engine) Written by Way of Letter to the Right Honorable Charles Lord Vicount of Dungarvan, Eldest Son to the Earl of Corke* (Oxford: Printed by H: Hall, printer to the University, for Tho: Robinson, 1660), 328–29, Early English Books Online.
71. Shapin and Schaffer, *Leviathan and the Air-Pump,* 30–31. The experiment was immortalized in Joseph Wright's famous painting *An Experiment on the Bird in the Air Pump* (1768), currently hanging at the National Gallery in London, in which a half-dozen candlelit faces surround the expiring bird, mesmerized by it and by the haggard, luminous face of the natural philosopher who manipulates life and death with the turn of a valve.
72. Dear, "Totius in Verba," 145.
73. Hall and Hall, eds., *Correspondence of Henry Oldenburg,* 427; Jones, *Ancients and Moderns: A Study of the Rise of the Scientific Movement in Seventeenth-Century England,* 262.
74. Spedding, Ellis, and Heath, eds., *Works of Francis Bacon,* 47.
75. Hooke, *Micrographia,* 1.
76. Wren, *Parentalia,* 204; Boyle, *The Excellency of Theology,* 71.
77. Wren, *Parentalia,* 223.
78. Robert Boyle, *Occasional Reflections* (London: Printed by W. Wilson for Henry Herringman, 1665), 19.
79. Lorraine J. Daston, "Marvelous Facts and Miraculous Evidence in Early Modern Europe," *Critical Inquiry* 18 (1991), 109–120.
80. Golinski, *"A Noble Spectacle"*; Hunter, *Establishing the New Science: The Experience of the Early Royal Society;* Shapin, *A Social History of Truth;* Shapin and Schaffer, *Leviathan and the Air-Pump.*

81. Birch, *Journal Books of the Royal Society*: Volume 1, 24 June 1663. The minutes of several meetings in 1663, soon after the Society received its royal charter, mentioned a series of such demonstrations that were to be brought together to impress Charles II (Council Minutes Original, Vol. 1, 1663–1681, MS CMO/1. Royal Society of London); it is not recorded whether the exposition ever happened, but we know the Society made presents to the king of some of their more spectacular results for his curiosity cabinet; see Golinski, "A Noble Spectacle: Phosphorus and the Public Cultures of Science in the Early Royal Society," 25.
82. Shapin and Schaffer, *Leviathan and the Air-Pump*, 77.
83. In the preface to the *Micrographia*, 6g, Hooke cites subscriptions from merchants as evidence that the Society is productively engaged in producing goods for the polity.
84. Jordynn Jack, "A Pedagogy of Sight: Microscopic Vision in Robert Hooke's Micrographia," *Quarterly Journal of Speech* 95, no. 2 (2009): 200–201.
85. Hooke, *Micrographia*, 4.
86. Samuel Butler, *Poetical Works of Samuel Butler* (London: William Pickering, 1835), 159.
87. Shapin, *A Social History of Truth*, 127. Some illustrative examples of Boylean hagiography during and immediately following his lifetime can be found in Sprat's *History* and Burnet, "A Sermon Preached at the Funeral of the Hon. Robert Boyle."
88. Robert Boyle, "An Account of Philaretus During His Minority," in *The Works of the Honourable Robert Boyle, in Five Volumes; to Which Is Prefixed the Life of the Author* (London: Printed for A. Millar, opposite Catharine-Street, in the Strand, 1744), 12. Other fellows' treatments of natural philosophy as a holy vocation include Charleton, *The Darkness of Atheism Dispelled by the Light of Nature. A Physico-Theological Treatise*, 167; Henry Oldenburg, "Introduction," *Philosophical Transactions of the Royal Society* 1(1665), http://0-rstl.royalsocietypublishing.org/; Wren, *Parentalia*, 221. *Founders' writings performed other key confirming signs such as asceticism—Beale*, Letter, 13 July 1667; Joseph Glanvill, Letter to Henry Oldenburg, 15 August 1669, folder EL/G1/16, MS Early Letters, Royal Society of London—and compulsion to work or share their work: Evelyn, *Letter to Robert Boyle*, September 1659; Glanvill, *Plus Ultra*, A2.
89. Henry Stubbe takes pains to cut Joseph Glanvill out of Boyle's flock before attacking him in Henry Stubbe, *Legends No Histories: Or, a Specimen of Some Animadversions Upon the History of the Royal Society. Wherein, Besides the Several Errors against Common Literature, Sundry Mistakes About the Making of Salt-Petre and Gun-Powder Are Detected, and Rectified: Whereunto Are Added Two Discourses, One of Pietro Sardi, and Another of Nicolas Tartaglia Relating to That Subject. Translated out of Italian. With a Brief Account of Those Passages of the Authors Life,*

Which the Virtuosi Intended Most to Censure, and Expatiate Upon: Written to Save Them the Trouble of Doing Any Thing Besides Defending Themselves. Together with the Plus Ultra of Mr. Ioseph Glanvill Reduced to a Non-Plus, &C. (London, 1670), 5, Early English Books Online.

90. Glanvill, *Plus Ultra*, 86–88; Hooke, Proposalls for the Good of the Royal Society; the "Dedicatory" of Wilkins, *An Essay Towards a Real Character, and a Philosophical Language*; Wren, *Parentalia*, 221.
91. Bacon, *Works*, 4:47.
92. Although it would be hard to find a more prophetic and Baconian document in the archive, I am excluding Sprat's *History of the Royal Society* from this discussion because, as mentioned previously, historians believe it was more a product of Sprat's idiosyncratic Baconianism than reflective of wider rhetorical practice in the Society, and because its goal was commemoration rather than dialogue.
93. To survey the major landmarks in this area: Charles Bazerman in *Shaping Written Knowledge* made a careful survey of the genre over the 200-year span from the first volume in 1665 to the mid-nineteenth century, when the form of the experimental report gelled more or less into the IMRD (Introduction, Method, Results, and Discussion) form still used today. Gross, Harmon, and Reidy zeroed in on the seventeenth century, the most volatile period rhetorically speaking, to study how and why narrative modes of experimental report were gradually squeezed out in favor of those that foregrounded hypothesis-testing; see their "Argument and 17th-Century Science: A Rhetorical Analysis with Sociological Implications," *Social Studies of Science* 30, no. 3 (2000): 371–96. Dwight Atkinson, in *Scientific Discourse in Sociohistorical Context* (Mahwah, NJ: Lawrence Erlbaum, 1999), has performed a linguistic analysis of the *PTRS* over 300 years, noting shifts from author-centric, gentlemanly discourse in the early years to a focus on objective methods and finally, in the twentieth century, to an emphasis on theory.
94. Gross, Harmon, and Reidy, "Argument and 17th-Century Science: A Rhetorical Analysis with Sociological Implications," 372.
95. Atkinson, *Scientific Discourse in Sociohistorical Context*, 98–101.
96. Golinski, "A Noble Spectacle," 16–17.
97. (1:53–55).
98. (1:127)
99. Shapin, *A Social History of Truth*, 191.
100. Shapin and Schaffer, *Leviathan and the Air-Pump*, 327.
101. Dear, *"Totius in Verba,"* 156.
102. (1:7–8).
103. (1:415–416); Sprat, *History of the Royal Society*, 158.
104. *PTRS* 1:125.

105. Golinski, "A Noble Spectacle," 28–29.
106. Stubbe, *Legends No Histories*, 38–39.
107. Stubbe, *Legends No Histories*, 97. Golinski, "A Noble Spectacle," 37.
108. Shapin and Schaffer, *Leviathan and the Air-Pump*, 113–203.
109. Despite Bacon's stated ideal of an open public society for the advancement of natural knowledge, and despite formal declarations of the same by the founders, Hobbes's charges of secrecy were well-founded. Fellowship to the Society was sharply restricted, and the monthly meetings in its first few years were not open to the public. Hooke argued strenuously in his Proposalls for the Good of the Royal Society (1673) that the library be kept off-limits to non-fellows, and that the Society jealously guard new technologies until patents were secure.This tug-of-war between publicity and privacy can be explained by the interaction of the prophetic function of Confirmation. Special access must be kept special. If every tradesman in the quarter could replicate Hooke's watch or Boyle's air pump, these men would lose their cachet, and the public would be confused about why they should listen to their natural philosophy instead of to their local midwife, surgeon, or doomsday prophet. We see the same confusion and argumentation today over the aura of verisimilitude that the Internet gives to reports of science news; see Lynda Walsh, *Sins against Science: The Scientific Media Hoaxes of Poe, Twain, and Others* (Albany, NY: SUNY Press, 2006), 215. At the same time, regular demonstrations of special access are required to maintain cachet (a word that itself stems from the word for "hidden"). The same conundrum was already present in Bacon's *New Atlantis* where Solomon's House kept its secrets until they feel the public is "ready" to know or use them. For all his trumpeting of a "public" society, Bacon never intended for all the workings of his College of the Six Days' Work to be utterly transparent. He was, as were all of his contemporaries, fond of ciphers for filtering out the uninitiated from the audience. Thus, we see the conflict between "plain style" and the cryptographic systems that the founders toyed with in the early years; cf. Bacon, *Works*, 4.254–55 and Wilkins, *An Essay Towards a Real Character.*
110. In his response to Boyle's *New Experiments Physico-Mechanical* (1660) detailing the operating of the air-pump, Hobbes published his *Dialogus Physicus* (1661), which has been translated by Simon Schaffer in *Leviathan and the Air-Pump*, 345–91. In the dialogue, a Hobbes stand-in, identified as "*A*," interrogates "*B*," a purported member of the Gresham Circle, about the experiments published in Boyle's book, among others. *A*'s first line of attack is the secrecy of the circle. Then, after *B* describes the spectacular killing of mice in the air-pump's receptacle by the evacuation of air, *A* protests that their inductive method is circular (353). He goes on to rebut Boyle's theory of the "spring of the air" as a cause of the vacuum claimed to reside in the air-pump receptacle with his own Aristotelian, plenist theory of air as a fluid. At the end, *A* claims the Society's experiments are too abstruse to be applied universally, which *A* believes is the test of a true philosophical principle.

111. Joseph Glanvill, *A Praefatory Answer to Mr. Henry Stubbe, the Doctor of Warwick Wherein the Malignity, Hypocrisie, Falshood of His Temper, Pretences, Reports, and the Impertinency of His Arguings & Quotations in His Animadversions on Plus Ultra Are Discovered*, (London: Printed by A. Clark, for I. Collins, at the Kings Arms in Ludgate-street, near the West-end of S. Pauls; and at the Kings Head in Westminster-hall, 1671), 29, Early English Books Online. Glanvill paints Stubbe as a raving false prophet: "He clamours, that *this Institution* is destructive to *Monarchy*, *Church* of *England*, *Universities*, and all *ancient Literature*; and follows his confident accusations with loud *outcries*, and restless *importunities*, and numerous *reproaches*; persecuting that illustrious Company with such *wild ravings*, and *impetuous eagerness*, as if it were an Army of *Painims* that had invaded us, and *he* the only man that saw the danger," ibid., 2.
112. Ibid., 47.
113. Shapin and Schaffer, *Leviathan and the Air-Pump*, 69.
114. Ibid., 343.

CHAPTER 5

1. Walsh, *Sins against Science*.
2. These studies include Bruno Latour's meditations on the social construction of pasteurization in *The Pasteurization of France*, trans. Alan Sheridan and John Law (Cambridge, MA: Harvard University Press, 1993); Greg Myers, "Nineteenth-Century Popularizations of Thermodynamics and the Rhetoric of Social Prophecy," *Victorian Studies* 29, no. 1 (1985): 35–66; John Angus Campbell's multiple studies of the transformative rhetoric of Charles Darwin beginning with "Darwin and the Origin of Species: The Rhetorical Ancestry of an Idea," *Speech Monographs* 37, no. 1 (1970): 1–14; and Thomas Lessl's examination of Thomas Huxley's priestly ethos in *Rhetorical Darwinism*.
3. The progressive model is sometimes called the linear model; see Pielke, *The Honest Broker: Making Sense of Science in Policy and Politics*, 12. For arguments that science was the muse of the Enlightenment, see Michael Adas, *Machines as the Measure of Men: Science, Technology, and Ideologies of Western Dominance* (Ithaca, NY: Cornell University Press, 1989); Creese, "The Scientist in Politics"; Ferris, *The Science of Liberty*; Don K. Price, "The Scientific Establishment," in *Scientists and National Policy-Making*, ed. Robert Gilpin and Christopher Wright (New York: Columbia University Press, 1964), 19–40; Paul Rabinow, *Anthropos Today: Reflections on Modern Equipment, Information Series* (Princeton, NJ: Princeton University Press, 2003); Wood, "Scientists and Politics: The Rise of an Apolitical Elite." For the Newton quote, see J. Robert Oppenheimer, *Science and the Common Understanding: The BBC Reith Lectures* (London: Oxford University Press, 1953), 4.
4. Simon LaPlace, *A Philosophical Essay on Probabilities* (New York: J. Wiley, 1814, 1902), 4, http://www.archive.org/details/philosophicaless00lapliala.

5. Immanuel Kant, "What Is Enlightenment?" in *Kant: On History*, ed. Louis White Beck (Indianapolis, IN: Library of Liberal Arts, 1784, 1963), 3.
6. David Hume, *Treatise of Human Nature* (Project Gutenberg, 1740, 2010), III.I.I, http://www.gutenberg.org/files/4705/4705-h/4705-h.htm#2H_4_0085.
7. Adas, *Machines as the Measure of Men*, 379.
8. Latour, *We Have Never Been Modern*.
9. The stasis of Action was called Procedure in the classical tradition, a term that invoked questions of jurisdiction. I use the term Action because this is the term adopted by most contemporary composition texts; however, note that this stasis can still invoke questions concerning the proper authority to handle a case. Though there are stasis precursors in Aristotle, Hermagoras is widely considered the originator of the classical theory. His key treatise (ca. 150 B.C.E.) has been lost, but the anonymous *Rhetorica ad Herennium* preserved Hermagoras's system. Cicero and Quintilian developed variations, but Hermogenes produced the dominant version of the classical stasis tradition on which medieval and later traditions were based: see Malcolm Heath, ed. *Hermogenes on Issues: Strategies of Argument in Later Greek Rhetoric* (Oxford, UK: Oxford University Press, 1995), 60–61.
10. Jeanne Fahnestock and Marie Secor, "The Stases in Scientific and Literary Argument," *Written Communication* 5 (1998): 427–43. Fahnestock and Secor were aware that cause/effect arguments were implied by the classical stasis of definition; see Michael Carter, "Stasis and Kairos: Principles of Social Construction in Classical Rhetoric," *Rhetoric Review* 7, no. 1 (1988): 97–112. Nevertheless, Fahnestock and Secor found that cause/effect arguments were foregrounded by the modern scientific method, which is in fact a system for developing consensus on the causes and effects of natural phenomena. So, they expanded the four-stasis system to a five-stasis system to account for the emphasis on cause/effect in scientific argumentation.
11. Hume, *Treatise of Human Nature*, III.i.i. para 12.
12. Jill Wrapson, "Artificial Fluoridation of Public Water Supplies in New Zealand: 'Magic Bullet,' 'Rat Poison,' or Communist Plot?" *Health & History* 7, no. 2 (2005): 17–29.
13. ABC News, "British Researcher Wakefield Defends Link between Vaccine and Autism," dir. George Stephanopoulos, 2011, http://abcnews.go.com/Health/Autism/autism-vaccine-link-research-dr-andrew-wakefield-fraud/story?id=12630566.
14. L. Trotter and S. Chapman, "'Conclusions about Exposure to ETS and Health That Will Be Unhelpful to Us': How the Tobacco Industry Attempted to Delay and Discredit the 1997 Australian National Health and Medical Research Council Report on Passive Smoking," *Tobacco Control* 12, suppl. 3 (2003): 102–106.
15. Weber, "Science as a Vocation," 143.

16. Albert Einstein, *Out of My Later Years* (New York: Philosophical Press, 1950), 22.
17. Merton, *Sociology of Science: Theoretical and Empirical Investigations*, 270–77.
18. Ian I. Mitroff, "Norms and Counter-Norms in a Select Group of the Apollo Moon Scientists: A Case Study of the Ambivalence of Scientists," *American Sociological Review 39*, no. 4 (1974): 579–95; Lawrence J. Prelli, "The Rhetorical Construction of Scientific Ethos," in *Landmark Studies in the Rhetoric of Science*, ed. Randy A. Harris (Mahwah, NJ: Lawrence Erlbaum Associates, 1997), 87–104; Merton, *Sociology of Science: Theoretical and Empirical Investigations*.
19. See technocratic essays on the characteristics of scientists: Julian Huxley, *Religion without Revelation*, (London: Max Parrish and Co., 1928, 1957), 236. See reviews of *Arrowsmith*, both the 1925 novel and the 1931 movie, in Henry Longan Stuart, "Lewis Assails Our Medicine Men," *New York Times*, March 8, 1925, News—Historical New York Times; Mordaunt Hall, "Arrowsmith (1931) the Screen; a Nobel Prize Novel," *New York Times* (December 8, 1931), http://movies.nytimes.com; Julian Huxley, *Religion without Revelation; and Thorstein Veblen's The Place of Science in Modern Civilization* (New York: Cosimo, Inc., 1919, 2007). Also, see contemporary media accounts quoted in Glen Scott Allen, "Master Mechanics & Evil Wizards: Science and the American Imagination from Frankenstein to Sputnik," *Massachusetts Review* 33 (1992): 505. Todd Wider in "The Positive Image of the Physician in American Cinema During the 1930s," *Journal of Popular Film & Television* 17, no. 4 (1990): 139 counts over two dozen films and serials in the 1930s portraying medical scientists; most cast them as saintly heroes, but some produce evils as a result of their blinkered devotion. This counter-image is developed in the character of the "mad scientist" that dominated horror films in the 1930s the most famous of which is of course Frankenstein; see Andrew Tudor, "Seeing the Worst Side of Science," *Nature 340*, no. 6235 (1989): 589–92.
20. Robert Henry Thurston, "Scientific Research: The Art of Revelation and of Prophecy," *Science* 16, September 12–19 (1902): 401–4, 454–57.
21. Beverly H. Burris, *Technocracy at Work* (Albany, NY: SUNY Press, 1993); Frank Fischer, *Technocracy and the Politics of Expertise* (London: Sage Publications, 1990).
22. Dwight D. Eisenhower, "Farewell Radio and Television Address to the American People," The American Presidency Project, University of California, Santa Barbara, originally delivered January 17, 1961, accessed September 8, 2011, http://www.presidency.ucsb.edu/ws/?pid=12086.
23. Theodore Porter, *Trust in Numbers: The Pursuit of Objectivity in Science and in Life* (Princeton, NJ: Princeton University Press, 1995), 1–8.
24. Weber, *Max Weber: The Theory of Social and Economic Organization*, 328–31.
25. Peter Galison, "Judgment against Objectivity," in *Picturing Science, Producing Art*, ed. Caroline Jones and Peter Galison (New York: Routledge, 1998), 353.

26. Porter, *Trust in Numbers*, 8.
27. Galison, "Judgment against Objectivity," 353.
28. Porter, *Trust in Numbers*, 7.
29. Ibid., 7.
30. Thorstein Veblen, *The Engineers and the Price System* (B. W. Huebsch, 1921), 143.
31. Howard Scott, "Hotel Pierre Address," Technocracy.org, accessed September 8, 2011, http://www.technocracy.org/technical-alliancetn/howard-scott/208-so-cietyscott; Veblen, The Engineers and the Price System.
32. Bush, "Science the Endless Frontier"; Howard Scott, "Cleveland Radio Address," Technocracy.org, accessed September 8, 2011, http://www.technocracy.org/technical-alliancetn/howard-scott/264-origns-scott.
33. Huxley, *Religion without Revelation.*
34. Merton, *Sociology of Science: Theoretical and Empirical Investigations*, 227–50.
35. Weber, "Science as a Vocation," 153, 46.
36. Ibid., 137–52.
37. Ibid., 137, 45–46.
38. Einstein, *Out of My Later Years*, 114.
39. Ferris, *The Science of Liberty*, 191; Rosslyn D. Haynes, *From Faust to Strangelove: Representations of the Scientist in Western Literature* (Baltimore, MD: Johns Hopkins University Press, 1994), 185; Lapp, *The New Priesthood*, 185; C.F. Weizsäcker, L. Castell, and O. Ischebeck, *Time, Quantum, and Information* (Springer, 2003), 28.
40. Allen, "Master Mechanics & Evil Wizards"; Burris, *Technocracy at Work*; Wider, "The Positive Image of the Physician in American Cinema During the 1930s."
41. Merton, *Sociology of Science: Theoretical and Empirical Investigations*, 261.
42. Haynes, *From Faust to Strangelove: Representations of the Scientist in Western Literature*, 88–89.
43. See Beard and Beard, *The American Leviathan*, p. 13, for the technology anxiety argument.
44. Cecelia Tichi, *Shifting Gears: Technology, Literature, Culture in Modernist America* (Chapel Hill, NC: University of North Carolina Press, 1996).
45. Daniel Dubarle, "The Scientist and His Responsibilities," *Bulletin of the Atomic Scientists 12* (1956): 254.
46. Contemporary sources for postwar language comparing scientists to thaumaturgs, priests, martyrs: Press Clippings, 1954, box 208, folder 4, J. Robert Oppenheimer Papers, Library of Congress; "Oedipus at Los Alamos," clipping from *Time* Magazine, November 2, 1959, box 200, folder 2–3, JRO Papers; "Fame by Association," clipping from *Newsweek*, August 30, 1965, box 200, folder 2–3, JRO Papers; "Eternal Apprentice," *Time*, 1948; Joseph Alsop and Stewart Alsop, "We Accuse!," *Harper's Magazine*, October 1954, 25–45; Kay Arnoll, "A New Kind of Hero," *New Outlook*, May 1954, 16–20, 24; Robert. E. Fitch, "Scapegoat

of the U.S. Purge," *New Leader*, September 26, 1955, 23–24; Lewis Gannett, "Book Review of Curtis's *The Oppenheimer Case*," *New York Herald Tribune*, April 30, 1956; Wayne A. R. Leys, "The Political Philosophy of Scientists," *Bulletin of the Atomic Scientists* 7, no. 1 (1951): 13–14; Price, "*The Scientific Establishment*"; Arthur M. Schlesinger Jr., "The Oppenheimer Case," *Atlantic Monthly*, October 1954, 29–36; Jacob J. Weinstein, "Strauss and Oppenheimer—a Contrast," K.A.M. News, March 18, 1955, box 213, folder 8, JRO Papers; Theodore H. White, "U.S. Science: The Troubled Quest—II," *Reporter*, September 7, 1954, 26–34; Albert Wohlstetter, "Strategy and the Natural Scientists," in *Scientists and National Policy-Making*, ed. Robert Gilpin and Christopher Wright (New York: Columbia University Press, 1964), 174–239; Wood, "Scientists and Politics: The Rise of an Apolitical Elite"; also, see quotations in Kai Bird and Martin Sherwin, *American Prometheus* (New York: Vintage Books, 2006), 323; Haynes, *From Faust to Strangelove: Representations of the Scientist in Western Literature*, 257. Representative utopian arguments at this time: Bush, "Science the Endless Frontier"; Oppenheimer, *Science and the Common Understanding: The BBC Reith Lectures;* Creese, "The Scientist in Politics"; Isidor I. Rabi, *Science and the Humanities*, December 6, 1955, box 197, folder 5, JRO Papers; also, see literary utopias quoted in Haynes, *From Faust to Strangelove: Representations of the Scientist in Western Literature*, 176–77.

47. Haynes, *From Faust to Strangelove: Representations of the Scientist in Western Literature*, 176; Tudor, "Seeing the Worst Side of Science."
48. Lapp, *The New Priesthood*, 128.
49. "Appendix D: Underground Nuclear Testing," *Nuclear Matters*, Office of the Deputy Assistant to the Secretary of Defense for Nuclear Matters, last modified 2008, http://www.acq.osd.mil/ncbdp/nm/nmbook/appendices/ap_D.htm, accessed September 11, 2011.
50. Karl R. Popper, *The Myth of Framework: In Defence of Science and Rationality* (New York: Routledge, 1994), 121–29.
51. Creese, "The Scientist in Politics," 4; Lapp, *The New Priesthood*, 9, 153–54; J. Robert Oppenheimer, "In the Keeping of Unreason" (paper presented at the Congress for Cultural Freedom, Rheinfelden, Switzerland, September 20–27, 1959); Wood, "Scientists and Politics: The Rise of an Apolitical Elite," 43.
52. Price, "The Scientific Establishment," 20.
53. Brooks, "The Scientific Adviser," 73.
54. Alvin M. Weinberg, "Impact of Large-Scale Science on the United States," *Science* 134, no. 3473 (1961): 161.
55. Francis Bello, "The Young Scientists," *Fortune*, June 1954, 142ff.; White, "U.S. Science: The Troubled Quest—II."
56. Brooks, "The Scientific Adviser," 86; Lapp, *The New Priesthood*, 203; McReynolds, "Note on the Compleat Expert," 15.
57. Brooks, "The Scientific Adviser," 85.

CHAPTER 6

1. Bird and Sherwin, *American Prometheus,* 324; J. Robert Oppenheimer, *Letter to Lewis L. Strauss, Chairman Atomic Energy Commission,* December 22, 1953, box 197, folder 4, JRO Papers.
2. Bird and Sherwin, *American Prometheus,* 179.
3. United States Atomic Energy Commission (AEC), *In the Matter of J. Robert Oppenheimer: Transcript of Hearing before Personnel Security Board* (Washington, DC: United States Government Printing Office, 1954), 461.
4. "Eternal Apprentice," 8.
5. Bird and Sherwin, *American Prometheus,* 317.
6. Ibid., 329.
7. J. Robert Oppenheimer, Statement Prepared for Special Senate Committee on Atomic Energy, December 6, 1945, box 237, folder 2, JRO Papers.
8. He invoked secrecy both in executive sessions and in public speeches, nearly always in an unprompted fashion (i.e., he did not wait for questions whose answers would compromise his security clearance). By asserting voluntarily that there were things he could not say, Oppenheimer injected into his rhetoric confirming reminders of his special access. Some representative examples follow. From his testimony before the Gray Board: "I think I need to point out that a great deal that happened between '45 and '49—I am not supposed to say to what extent—but to a very, very massive extent, we had become armed atomically." *In the Matter of J. Robert Oppenheimer,* 236. In the version of his "candor" speech published in the *Bulletin of Atomic Scientists*: "It is easy to say 'let us look at the arms race.' I must tell about it without communicating anything. I must reveal its nature without revealing anything." "Atomic Weapons and American Policy," *Bulletin of Atomic Scientists* 9, no. 6 (1953): 202.
9. Bird and Sherwin, *American Prometheus,* 373.
10. "Science: Expiation," *Time,* 1948, 94. Some other famous Oppie aphorisms: "Man is going to have to learn to live without war" from "Conversation with Physicists Robert Oppenheimer and Niels Bohr" (audio recording for *Tree of Knowledge/Harper's Monthly,* June 21, 1958), RWD 4782, Library of Congress; "We [the U.S. and the U.S.S.R.] may be likened to two scorpions in a bottle, each capable of killing the other, but only at the risk of his own life," from Oppenheimer, "Atomic Weapons and American Policy," 203; "[T]he best way to send information is to wrap it up in a person," from "Eternal Apprentice," 9; and when asked about the atomic bomb's "limitations," "The limitations lie in the fact that you don't want to be on the receiving end," "Eternal Apprentice," 8.
11. Reception of Fermi Award, 1963, box 226, folder 14, JRO Papers; Press Clippings, JRO Papers; "The Insecurity Risk," *Harper's Magazine,* October 1954, 14–17; Alsop and Alsop, "We Accuse!"; Arnoll, "A New Kind of Hero"; Schlesinger,

"The Oppenheimer Case"; "Weinstein, Strauss and Oppenheimer—a Contrast," K.A.M. News, JRO Papers.

12. Bird and Sherwin, *American Prometheus,* 314, 323–26.
13. Ibid., 390.
14. Richard Rhodes, *Dark Sun: The Making of the Hydrogen Bomb* (Simon & Schuster, 1996), 559.
15. Bird and Sherwin, *American Prometheus,* 532.
16. Ibid., 362.
17. Ibid., 465; J. Robert Oppenheimer, "Atomic Weapons and American Policy," *Foreign Affairs* 31, no. 4 (1953): 203.
18. Joseph Alsop and Stewart Alsop, "Operation Candor," *Herald Tribune,* 1953.
19. Bird and Sherwin, *American Prometheus,* 465–66.
20. Alsop and Alsop, "We Accuse!" 36; Bird and Sherwin, *American Prometheus,* 471.
21. Bird and Sherwin, *American Prometheus,* 546–50.
22. Dwight D. Eisenhower, *Top Secret to Herbert Brownell, Jr., 3 December 1953* (Baltimore, MD: Johns Hopkins University Press, 1996), http://www.eisenhowermemorial.org/presidential-papers/first-term/documents/583.cfm.
23. Gordon Gray and Thomas A. Morgan, *Findings and Recommendation of the Personnel Security Board in the Case of Dr. J. Robert Oppenheimer,* United States Atomic Energy Commission, Washington, DC, May 27, 1954, 35.
24. Bird and Sherwin, *American Prometheus,* 537.
25. In fact, after publishing their biography of Oppenheimer in 2005, Bird and Sherwin have mounted a legal campaign to have the clearance revocation overturned and an apology extended to what is left of the Oppenheimer family.
26. Oppenheimer, *Letter to Lewis L. Strauss, Chairman Atomic Energy Commission,* JRO Papers.
27. K. D. Nichols, *Letter to J. Robert Oppenheimer,* December 23, 1953, box 197, folder 4. JRO Papers.
28. AEC, *In the Matter of J. Robert Oppenheimer,* 330–740.
29. "Scientists Support Oppenheimer," *Bulletin of the Atomic Scientists,* September (1954): 283, 286.
30. (Box 208, folders 2–7).
31. AEC, *In the Matter of J. Robert Oppenheimer,* 567.
32. Oppenheimer, *Biography: Prewar Period,* 1954, box 198, folder 4, JRO Papers
33. AEC, *In the Matter of J. Robert Oppenheimer,* 229.
34. Ibid., 470.
35. Ibid., 229.
36. Ibid.
37. Bird and Sherwin, *American Prometheus,* 521.
38. AEC, *In the Matter of J. Robert Oppenheimer,* 470.
39. Ibid., 409.
40. Ibid., 270.

41. Ibid., 368.
42. Ibid., 236.
43. Ibid., 229–60.
44. Ibid., 402–10.
45. Bird and Sherwin, *American Prometheus,* 443–44.
46. Ibid., 418.
47. Ibid., 532.
48. AEC, *In the Matter of J. Robert Oppenheimer,* 710.
49. Ibid., 726.
50. Gray and Morgan, *Findings and Recommendation,* 33.
51. Ibid., 30.
52. Ibid., 35.
53. Lewis L. Strauss, Eugene M. Zuckert, and Joseph Campbell, *Statement by the Atomic Energy Commission,* United States Atomic Energy Commission, Washington, DC, June 29, 1954, 21–22.
54. Although the Gray Board had promised witnesses that they were testifying confidentially, and had blocked them from speaking to the press on the same grounds, Strauss nonetheless published the transcript almost immediately upon the AEC's decision, redacted for references to classified atomic secrets. It was reviewed by many national newspapers and magazines.
55. Bird and Sherwin, *American Prometheus,* 547, 547n, 556.
56. Roscoe Drummond, "Was the H-Bomb Traitorously Delayed?" *New York Herald Tribune,* 1953.
57. Bird and Sherwin, *American Prometheus,* 558.
58. Oppenheimer, "Man's Right to Knowledge" (paper presented at the Man's Right to Knowledge and the Free Use Thereof: Columbia University Bicentennial, New York, December 26, 1954), 1.
59. Oppenheimer, introduction to radio show "The World We Live In," March 29, 1954, box 262, folder 8, JRO Papers.
60. Bird and Sherwin, *American Prometheus,* 574.
61. We can see working here once more the ironic proliferation of hybrids. The hybrids in this case are the atomic and other "secrets" created by the intersection of nature with scientific culture. Nature, however, is no respecter of national or cultural boundaries. Thus, attempts to make secrets purely national objects—to strip them of their universal foundations in nature—inevitably fail because these secrets can only be controlled by controlling the minds, and therefore the bodies, that house them. As the sanctuary of national science is purified of bodies whose patriotism is suspect, the nation loses the secrets they carry; worse, the bodies then lodge elsewhere, in other nations, and start making more secrets there. Thus insecurity proliferates in response to security measures—this is the conundrum Stern explicated in his review of the Oppenheimer trial.

62. Philip M. Stern, *The Oppenheimer Case: Security on Trial* (New York: Harper & Row, 1969), 472.
63. Ibid., 470.

CHAPTER 7

1. Linda Lear, *Rachel Carson: Witness for Nature* (New York: Henry Holt, 1988), 39.
2. Ibid., 40.
3. Rachel Carson, research correspondence, 1959–1964, box 42, folders 752–78; box 43, folders 779–815; box 44, folder writings, Carson Papers, Beinecke Library, Yale University; Lear, Rachel Carson: *Witness for Nature*, 370.
4. Lear, *Rachel Carson: Witness for Nature*, 328.
5. Priscilla Coit Murphy, *What a Book Can Do: The Publication and Reception of Silent Spring* (Cambridge, MA: University of Massachusetts Press, 2005), 29–32.
6. Lear, *Rachel Carson: Witness for Nature*, 408.
7. Ibid., 423; Murphy, *What a Book Can Do*, 167.
8. Lear, *Rachel Carson: Witness for Nature*, 450; Murphy, *What a Book Can Do*, 98, 175.
9. Murphy, *What a Book Can Do*, 16.
10. Ralph H. Lutts, "Chemical Fallout: Silent Spring, Radioactive Fallout, and the Environmental Movement," in *And No Birds Sing: Rhetorical Analyses of Rachel Carson's Silent Spring*, ed. Craig Waddell (Carbondale: Southern Illinois University Press, 2000), 17–41.
11. Lear, *Rachel Carson: Witness for Nature*, 426; Murphy, *What a Book Can Do*, 156–57.
12. Lear, *Rachel Carson: Witness for Nature*, 450; Murphy, *What a Book Can Do*, 475; President's Science Advisory Committee, *Use of Pesticides* (Washington, DC, 1963), 23; Zuoyue Wang, "Responding to Silent Spring: Scientists, Popular Science Communication, and Environmental Policy in the Kennedy Years," *Science Communication* 19, no. 4 (1997): 151–55.
13. She performed more than just these hallmarks of prophetic ethos, as is evident throughout this chapter. However, the special vision, prophetic battles, and incarnated risks were highlighted in her rhetorical strategies in *Silent Spring* and by the producers of the CBS program, probably because they were the most dramatic elements.
14. Quoted in Murphy, *What a Book Can Do*, 130.
15. Lear, *Rachel Carson: Witness for Nature*, 447.
16. Ibid., 424.
17. Rachel Carson, *Silent Spring*, 40th anniversary ed. (New York: Houghton Mifflin, 2002), 199.

18. Ecosystems were a preoccupation of all Carson's writings. In the preface to the second edition of *The Sea around Us*, she created a careful causal chain to demonstrated how the seafloor was composed and then showed how nuclear testing disrupted it; see "Preface to the Second Edition of *the Sea around Us*," in *Lost Woods: The Discovered Writing of Rachel Carson*, ed. Linda Lear (Boston, MA: Beacon Press, 1961, 1998), 105–9. In her field notebooks, over and over again, she focuses on a small feature of the marine environment and then uses the topos of part to whole to imagine its significance in the over ecosystem: for instance, she reflected on some sand dunes near her Maine cottage:

    I stood where a new land was being built out of the sea, and I came away deeply moved. Although our intelligence forbids the idea, I believe our deeply rooted attitude toward the creation of the earth and the evolution of living things is a feeling that it all took place in a time infinitely remote. Now I understood. Here, as if for the benefit of my puny human understanding, the processes of creation—of earth building—had been speeded up so that I could trace the change within the life of my own contemporaries. The changes that were going on before my eyes were part and parcel of the same processes that brought the first dry land emerging out of the ancient and primitive ocean; or that led the first living creatures step by step out of the sea into the perilous new world of earth.

    Above quote from "Four Fragments from Carson's Field Notebooks," in *Lost Woods*, 130. In a letter written shortly before she died, she described her joy in her first transcontinental plane flight, which gave her a bird's-eye view of geological systems: "A privilege to see it that way. I'm full of thoughts about water cycles and the role of water in sculpting those strange canyon lands of the west" Lear, *Rachel Carson: Witness for Nature*, 465.
19. Carson, *Silent Spring*, 189.
20. Ibid., 1.
21. Ibid., 3.
22. Rachel Carson, "Design for Nature Writing," in *Lost Woods*, 96; "The Real World around Us," in *Lost Woods: The Discovered Writing of Rachel Carson*, 162; Lear, *Rachel Carson: Witness for Nature*, 423.
23. Lear, *Rachel Carson: Witness for Nature*, 422.
24. Murphy, *What a Book Can Do*, 39.
25. Ibid., 117–21.
26. Carson, *Silent Spring*, 13.
27. Ibid., 297.
28. Rachel Carson, "Women's National Press Club Speech," in *Lost Woods*, 210.
29. Carson, *Silent Spring*, 209.
30. Carson, "Women's National Press Club Speech," 203.
31. William J. Darby, "Silence, Miss Carson," *Chemical and Engineering News*, October 1, 1962, 62.

32. Lear, *Rachel Carson: Witness for Nature,* 430; James Westman, *News and Views from the Bureau of Conservation and Environmental Science,* no. 1 (November 1962), quoted in Kim Groshong, "The Noisy Reception of Silent Spring," in *An Element of Controversy: The Life of Chlorine in Science, Medicine, Technology and War,* ed. Hasok Chang and Catherine Jackson (London: Oxford University Press, 2007), 370.
33. Thomas H. Jukes, "A Town in Harmony," undated clipping from *Chemical Week,* RG 359 box 152, folder "Life Sciences folder—Pesticides 1962–1963," p. 45, Records of the Office of Science and Technology, National Archives. The chemical company Monsanto published another parody of "Fable for Tomorrow" called "The Desolate Year," quoted in Lear, *Rachel Carson: Witness for Nature,* 431.
34. Linda Lear, introduction to *Silent Spring* (New York: Houghton Mifflin, 2002), xvii; Michael B. Smith, "'Silence, Miss Carson!' Science, Gender, and the Reception of Silent Spring," *Feminist Studies* 27, no. 3 (2001): 733–52.
35. Lear, *Rachel Carson: Witness for Nature,* 409.
36. Jukes, Letter to Jerome Wiesner, August 31, 1962, RG 359, box 152, folder "Life Sciences folder—Pesticides 1962–1963," p. 43, OST.
37. Ibid., 437, 463.
38. Murphy, *What a Book Can Do,* 44. There is no evidence that Carson's cancer was precipitated by exposure to environmental toxins; nevertheless, given that she made the pesticide-cancer connection explicitly in *Silent Spring,* she was concerned that others would make the connection in her case and falsely attribute her authorial passion to that motive.
39. Eric Sevareid and Jay McMullen, "The Silent Spring of Rachel Carson," *CBS Reports* (CBS, April 3, 1963).
40. Lear, *Rachel Carson: Witness for Nature,* 450; Murphy, *What a Book Can Do,* 126.
41. Farrington, *Philosophy of Francis Bacon,* 67–68.
42. Carson, *Silent Spring,* 13.
43. Lear, Rachel Carson: *Witness for Nature,* 460; Murphy, *What a Book Can Do,* 36.
44. Murphy, *What a Book Can Do,* 33, 51, 87. For a contemporary account of the strongly normative culture of science, see Michael Polanyi, "The Republic of Science: Its Political and Economic Theory," *Minerva* 1 (1962): 54–74. Polanyi gives too much credit to the "republic of science" for its ability to self-govern and self-correct without guidance from larger polities; nonetheless, his observations about the internal enforcement of disciplinary norms of scientific communication have been amply borne out in subsequent research.
45. Murphy, *What a Book Can Do,* 44.
46. Carson, *Silent Spring,* 277.
47. Murphy, *What a Book Can Do,* 177.
48. Lear, *Rachel Carson: Witness for Nature,* 419; Murphy, *What a Book Can Do,* 71.
49. Minutes of Meeting Held 3 PM, 30 July 1962 on Pesticides, 1 August 1962, RG 359 box 152, folder "Life Sciences folder—Pesticides 1962–1963," p. 70, OST.

50. Murphy, *What a Book Can Do*, 92–93.
51. See RG 359 box 152, folder "Life Sciences folder—Pesticides 1962–1963," OST.
52. President's Science Advisory Committee, Life Sciences Panel: "Working Paper on Pesticides" (first draft), 1962, box 251, folder "Life Science—Pesticides, vol. 1, April 1, 1963," p. 51, OST.
53. Life Sciences Panel: Hazards of Pesticides, 1963, box 251, folder "Life Science—Pesticides, vol. 1, 1 April 1963," OST.
54. Ibid.
55. Ibid.
56. Cf. wordings in Carson, *Silent Spring*, 25, 74, 113, 246, 278.
57. President's Science Advisory Committee, *Use of Pesticides*, 28.
58. Ibid., 23.
59. Hearing on Interagency Coordination in Environmental Hazards before the Subcommittee on Reorganization and International Organizations of the Committee on Government Operations, 88th Cong. 206, June 4, 1963, 206.
60. The quote is apocryphal; the version Sevareid quoted is this one, from Charles Stowe's 1911 account of a meeting he most likely did not attend: "So you're the little woman who wrote the book that made this great war! Sit down please." Charles Edward Stowe and Lyman Beecher Stowe, *Harriet Beecher Stowe: The Story of Her Life* (Boston: Houghton Mifflin, 1911), 202. In fact, comparisons between *Uncle Tom's Cabin* and *Silent Spring* were so frequent in the media reception that I stopped counting them.
61. Hearing on Interagency Coordination in Environmental Hazards, 206.
62. Ibid.
63. Ibid., 210.
64. Ibid., 246.
65. Ibid., 223.
66. Lear, *Rachel Carson: Witness for Nature*, 455.

CHAPTER 8

1. Carl Sagan and Ann Druyan, "The Cosmic Ocean," *Cosmos* (KCET, September 28, 1980).
2. John Brockman, *The Third Culture* (New York: Simon & Schuster, 1995), 17; the "third culture" is a reference to C. P. Snow's famous lament about the lack of rapprochement between the "two cultures" of scientists and humanists. See also Rae Goodell, *The Visible Scientists* (Boston: Little, Brown, 1975); Lessl, "Gnostic Scientism and the Prohibition of Questions"; K. Giberson and M. Artigas, *Oracles of Science: Celebrity Scientists versus God and Religion* (New York: Oxford University Press, 2006). In this chapter, I borrow Giberson and Artigas's epithet for the popularizers, but remember that in general I reserve

this term for prophetic institutions, preferring to call the people who constitute them prophets.

3. Giberson and Artigas, *Oracles of Science*, 4.
4. For a broad discussion of the effect of media on Sagan's ethos, see David Tietge, *Rational Rhetoric: The Role of Science in Popular Discourse* (West Lafayette, IN: Parlor Press, 2008); for an excellent but quite narrow structuralist analysis of the media text of *Cosmos*, see Roger Silverstone, "Narrative Structures in Television Science," in *Impacts and Influences: Essays on Media Power in the Twentieth Century*, ed. James Curran, Anthony Smith, and Pauline Wingate (New York: Methuen, 1987): 291–330.
5. Robert C. Cowen, "'Cosmos': Whether You Liked It on TV or Not, by All Means Read the Book," *Christian Science Monitor*, November 19, 1980, LexisNexis.
6. Daniel Dennett, *Darwin's Dangerous Idea: Evolution and the Meaning of Life* (New York: Simon & Schuster, 1996), 310.
7. Simon Jenkins, "A Dance to the Music of Imaginary Time; Stephen Hawking, Writer," *London Times*, August 28, 1988, LexisNexis.
8. See the "Dedication" to Stephen Jay Gould, *Ever Since Darwin: Reflections in Natural History* (New York: W. W. Norton, 1977); Louise Sweeney, "Carl Sagan: Reviving Our Sense of Wonder," *Christian Science Monitor*, March 18, 1982, LexisNexis.
9. Giberson and Artigas, *Oracles of Science*, 126.
10. Kitty Ferguson, *Stephen Hawking: Quest for a Theory of Everything* (New York: Bantam Books, 1991), 41–44.
11. Stephen Jay Gould, "Fulfilling the Spandrels of World and Mind," in *Understanding Scientific Prose*, ed. Jack Selzer (Madison: University of Wisconsin Press, 1993), 321.
12. Carl Sagan, *The Demon-Haunted World: Science as a Candle in the Dark* (New York: Random House, 1995), 30.
13. Stephen W. Hawking, "The Future of the Universe," in *Predicting the Future*, ed. Leo Howe and Alan Wain, 8–23 (Cambridge, UK: Cambridge University Press, 1993).
14. Despite the fact he sustained his scientific work at Cornell at some level throughout his popularizing career and was involved in the seminal Voyager program, Sagan was denied membership in the National Academy of Sciences in a contentious vote in 1992. His ex-wife, Lynn Margulis, who had been made an NAS fellow several years before for her groundbreaking work in eukaryotic biology, attended the vote and attributed the snub to his "outspoken attitude especially on nuclear winter." Quoted in Giberson and Artigas, *Oracles of Science*, 125. The NAS extended an olive branch a few years later to Sagan in the form of its Public Welfare Medal.
15. Goodell, *The Visible Scientists*, 178.

16. Sagan, *The Demon-Haunted World*, 4.
17. Judy Bachrach, "A Beautiful Mind, an Ugly Possibility," *Vanity Fair*, June 2004, http://www.vanityfair.com/culture/features/2004/06/hawking200406.
18. For samples of the argument attributing Hawking's fame in part to audience sympathy for his disability, see Charles Oulton, "Cosmic Writer Shames Book World; Professor Stephen Hawking," *Sunday Times*, August 28, 1988, LexisNexis; Tietge, *Rational Rhetoric*, 189.
19. Jeanne Fahnestock, "Accomodating Science: The Rhetorical Life of Scientific Facts," *Written Communication* 3, no. 3 (1986): 275–96; Carolyn G. Hartnett, "How Does Science Express Uncertainty," *LACUS Forum* 30 (2004): 355–65; Marcel LaFollette, "Science on Television," *Daedalus* 111, no. 4 (1982): 188; Silverstone, "Narrative Structures in Television Science," 305; Jay A. Winsten, "Science and the Media: The Boundaries of Truth," *Health Affairs* 4, no. 1 (1985): 6–9.
20. Andrew Brown, *The Darwin Wars: The Scientific Battle for the Soul of Man* (London: Simon & Schuster, 1999), 149.
21. Gould claimed that he wrote the first paper, while Eldredge supplied the ideas; future collaborations had a more even distribution of writing work. Stephen Jay Gould, "Opus 200," *Natural History* 100 (August 1991), 12–18.
22. Stephen Jay Gould and Niles Eldredge, "Punctuated Equilibrium: The Tempo and Mode of Evolution Reconsidered," *Paleobiology* 3, no. 2 (1977): 146; David Forrest Prindle, *Stephen Jay Gould and the Politics of Evolution* (Amherst, NY: Prometheus, 2009), 20.
23. Gould, "Fulfilling the Spandrels of World and Mind," 317.
24. Bruce Ward, "Dueling Darwinists: As Stephen Jay Gould Prepares His 'Big Book' on Evolutionary Biology, He Is Increasingly under Attack for His Theory of 'Punctuated Equilibrium.' His Critics Call Him a Fuzzy Evolutionary Thinker, Guilty of 'Bad Scientific Poetry,'" *Ottawa Citizen*, October 17, 1998, LexisNexis.
25. Dennett, *Darwin's Dangerous Idea*, 310.
26. John Alcock, "Unpunctuated Equilibirum in the *Natural History* Essays of Stephen Jay Gould," *Evolution and Human Behavior 19*, no. 5 (1998): 323.
27. Ibid., 326.
28. Brown, *The Darwin Wars*, 118.
29. "'Darwinian Fundamentalism': An Exchange," *New York Review of Books*, August 14, 1997, LexisNexis.
30. Gillian Fuller, "Cultivating Science: Negotiating Discourse in the Popular Texts of Stephen Jay Gould," in *Reading Science: Critical and Functional Perspectives on Discourses of Science*, ed. J. R. Martin and Robert Veel (Psychology Press, 1998), 40.
31. Boyce Rensberger, "Recent Studies Spark Revolution in Interpretation of Evolution," *New York Times*, November 4, 1980, LexisNexis.
32. Prindle, *Stephen Jay Gould and the Politics of Evolution*, 97.

33. Alcock, "Unpunctuated Equilibirum in the *Natural History* Essays of Stephen Jay Gould"; Brown, *The Darwin Wars*, 151; Dennett, *Darwin's Dangerous Idea*, 278; John Maynard Smith, "Genes, Memes & Minds," *New York Review of Books*, November 30, 1995.
34. Jane Gregory and Steven Miller, *Science in Public: Communication, Culture, and Credibility* (Plenum Trade, 1998), 130.
35. Silverstone, "Narrative Structures in Television Science," 302.
36. Gregory and Miller, *Science in Public*, 121; LaFollette, "Science on Television," 187.
37. Jordynn Jack, "Chronotopes: Forms of Time in Rhetorical Argument," *College English* 69, no. 1 (2006): 52–73.
38. Ken Baake, *Metaphor and Knowledge: The Challenges of Writing Science* (Albany, NY: SUNY Press, 2003), 71.
39. Silverstone, "Narrative Structures in Television Science," 317.
40. Ibid., 319.
41. The main criticism of *Cosmos* repeated in contemporary reviews, many of them written by other scientists or science journalists, was the way it singled Sagan out as a solitary hero when in fact he could not do his work without a large supporting community. One ascerbic scientist attending the gala screening of *On the Shores of the Cosmic Ocean* at the National Academy of Sciences was overheard expressing his hope that the next time one of the visionary trapdoors opened in the Ship of the Imagination, Sagan would fall through it; quoted in Tom Zito, "Carl Sagan, Cosmonaut," *Washington Post*, September 19, 1980, LexisNexis.
42. Sagan, *The Demon-Haunted World*, 29.
43. Bachrach, "A Beautiful Mind, an Ugly Possibility," 4.
44. Hawking, *A Brief History of Time: From the Big Bang to Black Holes* (New York: Bantam Books, 1988), 175.
45. There is a wonderful sequence in the movie where Morris asks several cosmologists to describe what it would be like to fly into a "black hole," and they describe a quasi-ecstatic ascetic process by which the mind would perceive processes moving faster and faster until they exceeded description, then comprehension, and then the perceiver's body would be stretched and stretched to the point where it would suddenly be "crushed into spaghetti"; quoted in Errol Morris, *A Brief History of Time*, (Paramount Home Video, 1992), 00:33–38. One of the physicists interviewed, Brandon Carter, opined that he would wish to "go" that way (00:39).
46. Morris, *A Brief History of Time*, 01:07–01:09.
47. Ken Baake, *Metaphor and Knowledge*, 52–72.
48. Ibid., 56.
49. Andrew Brown, "Altruistic Equations That Killed a Good Man," *Independent*, October 14, 1996, 14.

50. Walsh, *Sins against Science,* 21.
51. G. Hammond, "God Knows—but He's Not Telling," *Sunday Mail,* December 4, 1988, LexisNexis.
52. Silverstone, "Narrative Structures in Television Science," 306. Cf. the argument that photography is treated as transparent and, since videography is even more "realistic" than photography, this illusion is strengthened; in Gunther Kress and Theo Van Leeuwen, *Reading Images: The Grammar of Visual Design* (New York: Routledge, 1996, 2006), 26.
53. Carl Sagan and Ann Druyan, "Who Speaks for Earth?" *Cosmos* (KCET, December 21, 1981).
54. Goodell, *Visible Scientists,* 176.
55. Sagan and Druyan, "Who Speaks for Earth?" 00:49–50.
56. Sweeney, "Carl Sagan: Reviving Our Sense of Wonder."
57. For the reception of two such speeches in Canada, see Canadian Press, "Canadian Planetary Perspective Can Lower World Tensions: Sagan," *Globe and Mail,* June 5, 1984, LexisNexis; Norman Webster, "Final Absurdity Must Be Faced," *Globe and Mail,* May 22, 1982, LexisNexis.
58. Starley L. Thompson and Stephen H. Schneider, "Nuclear Winter Reappraised," *Foreign Affairs* 64, no. 5 (1986).
59. Sagan and Druyan, "Who Speaks for Earth?" epilogue.
60. Vincent Mosco, "Star Wars Is Already Working," *Science as Culture* 1, suppl. 001 (1987): 12–34; Rabinow in *Anthropos Today: Reflections on Modern Equipment,* 103, claims 75% of Cold War funding for science was military in aim.
61. Sweeney, "Carl Sagan: Reviving Our Sense of Wonder."
62. Sagan, *Demon-Haunted World,* 38.
63. Tietge, *Rational Rhetoric,* 125.
64. Richard Lewontin, "Science's Demons," *Australian,* March 22, 1997, LexisNexis.
65. Jon D. Miller, "The Measurement of Civic Scientific Literacy," *Public Understanding of Science* 7 (1998): 203–23.
66. John Allen Paulos, "The Odds Are You're Innumerate," *New York Times,* January 1, 1989, LexisNexis.
67. Tim Radford, "Book Thoughts: Master of the Universe," *Guardian,* December 30, 1988, LexisNexis.
68. Maureen Madison, "Is Stephen Hawking Modern? A Study of a Brief History of Time in Relation to the Theories of Bruno Latour," *Janus Head* (2001), http://www.janushead.org/gwu-2001/madison.cfm.
69. Richard Zoglin, "Science on TV—How Sharp Is the Focus?" *New York Times,* April 26, 1981, LexisNexis.
70. Michael Stewart, "Upwardly Mobile Phyla Facts," *Sunday Times,* February 11, 1990, LexisNexis.
71. Robert Wilson, "Gould, Chipping Away at the Foundation of Evolution," *USA Today,* November 3, 1989, LexisNexis.

72. Sebastian Faulks, "Deafened by the Silence from Science," *Independent,* March 4, 1990, LexisNexis.
73. Fuller, "Cultivating Science: Negotiating Discourse in the Popular Texts of Stephen Jay Gould," 59.
74. Stephen Jay Gould, *The Panda's Thumb* (New York: W. W. Norton, 1980), 313.
75. Ibid., 11.
76. Ibid., 313.
77. Davida Charney, "A Study in Rhetorical Reading: How Evolutionists Read 'The Spandrels of San Marco,'" in *Understanding Scientific Prose,* ed. Jack Selzer (Madison, WI: University of Wisconsin Press, 1993), 218–28.
78. Carolyn Miller, "The Presumptions of Expertise: The Role of Ethos in Risk Analysis," *Configurations: A Journal of Literature, Science, and Technology 11,* no. 2 (2003): 196–97.
79. These were first identified as "scientific values" in Polanyi, "The Republic of Science: Its Political and Economic Theory." See a related set in Patrick Grim, "Scientific and Other Values," in *Philosophy of Science and the Occult,* ed. Patrick Grim (Albany, NY: SUNY Press, 1982), 315.
80. Gould, "Fulfilling the Spandrels of World and Mind," 321.
81. Dennett, *Darwin's Dangerous Idea,* 263–64, 310.
82. Ibid., 264.
83. Ibid., 278.
84. Charney, "A Study in Rhetorical Reading," 220.
85. Heidi Scott, "Stephen Jay Gould and the Rhetoric of Evolutionary Theory," *Rhetoric Review 26,* no. 2 (2007): 126.
86. Maynard Smith, "Genes, Memes & Minds."
87. Fuller, "Cultivating Science: Negotiating Discourse in the Popular Texts of Stephen Jay Gould," 59.
88. Giberson and Artigas, *Oracles of Science,* 12.
89. Charles Bazerman, "Intertextual Self-Fashioning: Gould and Lewontin's Representations of the Literature," in *Understanding Scientific Prose,* ed. Jack Selzer (Madison, WI: University of Wisconsin Press, 1993), 38.
90. Prindle, *Stephen Jay Gould and the Politics of Evolution,* 199.
91. "Testimony of Dr. Stephen Jay Gould, Professor of Geology, Harvard University (Plaintiffs Witness)—Transcript Paragraph Formatted Version," *McLean v. Arkansas Documentation Project* (1981, 2005), http://www.antievolution.org/.
92. Stephen Jay Gould, *Hen's Teeth and Horse's Toes* (New York: W. W. Norton, 1994), 261.
93. Stephen Jay Gould, "Evolution as Fact and Theory," *Discover 2* (1981): 34–37.
94. Stephen Jay Gould, *Rocks of Ages: Science and Religion in the Fullness of Life* (New York: Random House, 1999), 3.
95. Ferguson, *Stephen Hawking: Quest for a Theory of Everything,* 42. Shortly after the publication of Ferguson's biography, the Hawkings divorced. Jane Hawking has

since published a memoir in which she makes it clear that the relentless work of providing for Stephen's needs contributed to the split.

96. The NSF currently estimates that about 1 in 7 physicists and astronomers are women, compared with a national average of 1 in 4 for all sciences; see National Science Foundation, "Table 9-5: Employed Scientists and Engineers, by Occupation, Highest Degree Level, and Sex: 2006," accessed February 7, 2012, http://www.nsf.gov/statistics/wmpd/pdf/tab9-5.pdf.

97. For a concise overview of the representation of women and minorities in the science, technology, engineering, and math (STEM) fields, and the particular challenges they face, see Roger J. Burke, "Women and Minorities in Stem: A Primer," in *Women and Minorities in Science, Technology, Engineering and Mathematics: Upping the Numbers,* ed. R. J. Burke and M. C. Mattis (Northhampton, MA: Edward Elgar, 2007), 3–27.

98. Eva Flicker, "Between Brains and Breasts; Women Scientists in Fiction Film: On the Marginalization and Sexualization of Scientific Competence," *Public Understanding of Science* 12, no. 3 (2003): 307–18; Kenneth King, "Educational Television: 'Let's Explore Science'," *Journal of Science Education and Technology* 9, no. 3 (2000) 227–46; LaFollette, "Science on Television"; LaFollette, "Eyes on the Stars: Images of Women Scientists in Popular Magazines," *Science, Technology, and Human Values* 13, no. 3&4 (1988): 262–75; Dorothy Nelkin, *Selling Science: How the Press Covers Science and Technology,* revised ed. (Chicago, IL: University of Chicago Press, 1995); Jocelyn Steinke and Marilee Long, "A Lab of Her Own?" *Science Communication 18,* no. 2 (1996): 91–115.

99. Judith White, "Hawking Unravels Cosmic Mysteries," *Sydney Sun Herald,* September 4, 1988, LexisNexis. For the Prometheus reference relating to Sagan and the Oracles, see Thomas M. Lessl, "Science and the Sacred Cosmos: The Ideological Rhetoric of Carl Sagan," *Quarterly Journal of Speech* 71 (1985): 175.

100. There are numerous instances of this kind of language in professional reviews of the works of all three Oracles. As an example of an extended version of this critique, James Gleick, a cultic prophet himself, dedicated a substantial section of his review of *Wonderful Life* to pointing out Gould's insider status and suggesting that it blinded him to his own biases in interpreting the Burgess Shale; see "Survival of the Luckiest," *New York Times,* October 22, 1989, LexisNexis.

101. As we have seen, Carson's advocacy led to the banning of DDT and, indirectly, to the formation of the Environmental Protection Agency. Brockovich helped victims of chromium (VI) poisoning win a massive settlement from Pacific Gas & Electric in 1996. Goodall's institute successfully lobbied Congress to pass the Great Apes Conservation Act of 2000, and Goodall's testimony was the centerpiece of the debate to reauthorize the Act and its fund in 2010. As for the Oracles, Hawking has not engaged in political activism; Sagan's activism had mixed results, none as directly observable as the Carson effect; Gould was instrumental in winning *McLean v. Arkansas,* a major civic act, but

at other times appeared to pride himself on his skill at avoiding community service—see Stephen Strauss, "To S. J. Gould, the Meaning of Life Is Luck from Panda Thumbs to Paleontology, This Expert on Everything Is Always Intriguing," *Globe and Mail*, January 20, 1990, LexisNexis. The argument here is not necessarily that the women scientist-prophets have been *more* effective than their male counterparts in starting activist dialogue, but rather that the women's civic engagement challenges the scholarly truism that women have been silenced by their marginalization from scientific power centers and TV science.

102. Sweeney, "Carl Sagan: Reviving Our Sense of Wonder."
103. Lessl, "Science and the Sacred Cosmos: The Ideological Rhetoric of Carl Sagan," 176; Tietge, *Rational Rhetoric*, 133–34.
104. Zoglin, "Science on TV—How Sharp Is the Focus?"
105. Andrew Jamison and Ron Eyerman, *Seeds of the Sixties* (Berkeley: University of California Press, 1994), 96–100; as a reminder, I'm taking my idea of hybridity from Bruno Latour's theory of modernity. Hybrids of nature and culture ironically proliferate as modern societies try to "purify" each category of influences from the other; see Latour, *We Have Never Been Modern*.
106. Gregory and Miller, *Science in Public: Communication, Culture, and Credibility*, 119.
107. Giberson and Artigas, *Oracles of Science*, 5.

CHAPTER 9

1. S. Fred Singer, "The End of the IPCC: One Mistake Too Many!" Nongovernmental International Panel on Climate Change, accessed December 22, 2011, http://www.nipcc.ch/datei/1268563724.pdf.
2. Paul N. Edwards, *A Vast Machine: Computer Models, Climate Data, and the Politics of Global Warming* (Cambridge, MA: MIT Press, 2010), 412; Naomi Oreskes and Erik M. Conway, *Merchants of Doubt: How a Handful of Scientists Obscured the Truth on Issues from Tobacco Smoke to Global Warming* (New York: Bloomsbury Press, 2010), 5–8.
3. Silvio O. Funtowicz and Jerome R. Ravetz, "Science for a Post-Normal Age," *Futures* 25 (1993): 739–55.
4. Daniel Sarewitz and Roger A. Pielke Jr., "Prediction in Science and Policy," in *Prediction: Science, Decision Making, and the Future of Nature*, ed. Daniel Sarewitz, Roger A. Pielke, Jr., and Radford Byerly, Jr. (Washington, DC: Island Books, 2000), *11*.
5. Edwards, *A Vast Machine*, 399.
6. Intergovernmental Panel on Climate Change, "History," United Nations Environment Program, accessed January 5, 2012, http://www.ipcc.ch/organization/organization_history.shtml.

7. Climate historian James Fleming argues that Svante Arrhenius and John Tyndall, often heralded as the "fathers" of global warming, meant something quite different than we mean today when they used terms such as "hot-house theory"; they did indeed posit a connection between CO_2 concentrations in the atmosphere and warming, but they were discussing the effects of latitude, not industry. Toward the end of his career in 1906, Arrhenius came around to considering industrial carbon emissions, but then he tended to look on them favorably as introducing a "more equable" climate for the earth's inhabitants, featuring milder winters and more vigorous plant growth. *Historical Perspectives on Climate Change* (New York: Oxford University Press, 1998), 79–82.
8. For arguments about 1988 being a watershed in public awareness of climate change, see Ferris, *The Science of Liberty;* Mike Hulme, *Why We Disagree about Climate Change* (Cambridge, UK: Cambridge University Press, 2009), 64. On Jim Hansen's moral and religious rhetoric, see James Hansen and Amy Goodman, "Leading Climate Scientist James Hansen on Why He's Pleased the Copenhagen Summit Failed, 'Cap and Fade,' Climategate and More," Democracy Now, accessed January 6, 2012, http://www.democracynow.org/; Elizabeth Kolbert, "The Catastrophist," *New Yorker* June 29, 2009: 45–46; Roger A. Pielke Jr., *The Climate Fix: What Scientists and Politicians Won't Tell You A\about Global Warming* (New York: Basic Books, 2010), 208. The title of Hansen's 2008 book was *Storms of My Grandchildren: The Truth about the Coming Climate Catastrophe and Our Last Chance to Save Humanity.*
9. In the United States, congressional hearings on global warming and related topics have averaged 10 a year since 1988; they averaged about two a year between 1975 and 1988. Hulme, *Why We Disagree about Climate Change*, 64. For an analysis of the relationship between media and political awareness of climate change in Germany since 1988 see Peter Weingart, Anita Engels, and Petra Pansegrau, "Risks of Communication: Discoures on Climate Change in Science, Politics, and the Mass Media," *Public Understanding of Science* 9 (2000): 261–83. Although the political response to climate change has been quite different between Germany and the United States, the overall issue-attention cycle has probably been quite similar here and there; see Sheila Jasanoff, "Cosmopolitan Knowledge: Climate Science and Global Civic Epistemology," in *The Oxford Handbook of Climate Change and Society*, ed. John S. Dryzek, Richard B. Norgaard, and David Schlosberg (Oxford, UK: Oxford University Press, 2011), 127–43.
10. Simon Shackley and Brian Wynne, "Representing Uncertainty in Global Climate Change Science and Policy: Boundary-Ordering Devices and Authority," *Science Technology Human Values 21*, no. 3 (1996): 281.
11. Jasanoff, "Cosmopolitan Knowledge: Climate Science and Global Civic Epistemology," 133; Miller, "The Presumptions of Expertise: The Role of Ethos in Risk Analysis," 165.

12. Richard D. Besel, "Communicating Climate Change: Climate Rhetorics and Discursive Tipping Points in U.S. Global Warming Science and Public Policy" (PhD diss., University of Illinois, Urbana-Champaign, 2007), 91–92; Hulme, *Why We Disagree about Climate Change,* 64.
13. Besel, "Communicating Climate Change," 195–96; Edwards, *Vast Machine,* 409.
14. While neither of these scholars explicitly applies the stases to their analyses, Susan Peck MacDonald has noted shifts from arguments of fact and definition toward cause/effect arguments and evaluations in attachment psychology as the field has grown; see *Professional Academic Writing in the Humanities and Social Sciences* (Carbondale: Southern Illinois University Press, 1994), 53–73. Both Atkinson and Gross, Harmon & Reidy noted related shifts from observations toward explanations over time in their discourse analysis of experimental reports from the seventeenth to the twentieth century; see Atkinson, *Scientific Discourse in Sociohistorical Context,* 98–101; Alan G. Gross, Joseph E. Harmon, and Michael S. Reidy, *Communicating Science: The Scientific Article from the 17th Century to the Present* (West Lafayette, IN: Parlor Press, 2002), 230.
15. *IPCC, Second Assessment Report: Climate Change 1995* (Cambridge, UK, and New York,: Cambridge University Press, 1995), 22, http://www.ipcc.ch/pdf/climate-changes-1995/ipcc-2nd-assessment/2nd-assessment-en.pdf.
16. Hulme, *Why We Disagree about Climate Change,* 38.
17. Edwards, *A Vast Machine,* 401. Peter Weingart has studied the kairoi of the unfolding climate change drama in Germany over this same period and termed the cycle he observed "cassandra syndrome": first, the media amplified a novel jeremiad issued by a few climate scientists; government reacted by allocating funds to the problem; the funding drove work that diversified and complicated the original predictions of doom; meanwhile, the interest-attention cycle for climate change in the media had spun down; so, the media was receptive to skeptics who seized on the diversity of climate science opinion to accuse the original jeremiahs of alarmism or bias; see "Science and the Media," *Research Policy* 27 (1998): 869–79. Weingart attributed this "cassandra syndrome" to the new interdependence of science and media. I would argue instead that the coupling merely amplified an old rhetorical catch-22 that is triggered whenever incompatible models for science advising cross at the stasis of cause/effect.
18. Ibid., 412, 429.
19. Ibid., 423–24.
20. Pielke, *The Climate Fix: What Scientists and Politicians Won't Tell You about Global Warming,* 200–11.
21. For technical uncertainties, see Silvio D. Funtowicz and Jerome R. Ravetz, *Uncertainty and Quality in Science for Policy* (Dordrecht: Kluwer Academic, 1990); M. G. Morgan and M. Henrion, *Uncertainty: A Guide to Dealing with Uncertainty in Quantitative Risk and Policy Analysis* (Cambridge, UK: Cambridge University

Press, 1990); William D. Rowe, "Understanding Uncertainty," *Risk Analysis* 14, no. 5 (1994): 743–50; Shackley and Wynne, "Representing Uncertainty in Global Climate Change Science and Policy"; Charles Weiss, "Expressing Scientific Uncertainty," *Law, Probability, and Risk* 2 (2003): 25–46; For epistemic uncertainty, see R. H. Moss and S. H. Schneider, "Uncertainties in the IPCC TAR: Recommendations to Lead Authors for More Consistent Assessment and Reporting," in *Guidance Papers on the Cross Cutting Issues of the Third Assessment Report of the IPCC*, ed. R. Pachauri, T. Taniguchi, and K. Tanaka (Geneva: World Meteorological Organization, 2000): 34–51; Michael J. Smithson, "Social Theories of Ignorance," in *Agnotology: The Cultural Production of Ignorance*, ed. R. Proctor and L. Schiebinger (Palo Alto, CA: Stanford University Press, 2008), 209–229. For linguistic uncertainty, see Fahnestock, "Accomodating Science"; Goodnight, "The Personal, Technical, and Public Spheres of Argumentation"; Hartnett, "How Does Science Express Uncertainty"; Ken Hyland, *Hedging in Scientific Research Articles* (Philadephia, PA: John Benjamins, 1998); Bruno Latour and Steve Woolgar, *Laboratory Life: The Construction of Scientific Facts* (Princeton, NJ: Princeton University Press, 1986). For rhetorical uncertainty, see Robert Danisch, "Political Rhetoric in a World Risk Society," *Rhetoric Society Quarterly* 40, no. 2 (2010): 172–92; Jeffrey Grabill and W. Michele Simmons, "Toward a Critical Rhetoric of Risk Communication: Producing Citizens and the Role of Technical Communication," *Technical Communication Quarterly* 7, no. 4 (1998): 415–41; David Michaels, *Doubt Is Their Product: How Industry's Assult on Science Threatens Your Health* (New York: Oxford University Press, 2008); Gordon R. Mitchell, *Strategic Deception: Rhetoric, Science, and Politics in Missile Defense Advocacy*, Michigan State University Press Rhetoric and Public Affairs Series (East Lansing: Michigan State University Press, 2000); Oreskes, *Merchants of Doubt*; Marcus Paroske, "Deliberating International Science Policy Controversies: Uncertainty and Aids in South Africa," *Quarterly Journal of Speech* 95, no. 2 (2009): 148–70; J. Blake Scott, "Kairos as Indeterminate Risk Management: The Pharmaceutical Industry's Response to Bioterrorism," *Quarterly Journal of Speech* 92, no. 2 (2006): 115–43; S. Holly Stocking and Lisa.W. Holstein, "Manufacturing Doubt: Journalists' Roles and the Construction of Ignorance in a Scientific Controversy," *Public Understanding of Science* 18 (2009): 23–42; Stephen C. Zehr, "Scientists' Representations of Uncertainty," in *Communicating Uncertainty: Media Coverage of New and Controversial Science*, ed. Sharon M. Friedman, Sharon Dunwoody, and Carol L. Rogers (Mahwah, NJ: Lawrence Erlbaum Associates, 1999), 3–22. Finally, for sociopolitical uncertainty, see Ulrich Beck, *World Risk Society* (Cambridge, UK: Polity Press, 1999); Brian L. Campbell, "Uncertainty as Symbolic Action in Disputes among Experts," *Social Studies of Science* 15, no. 3 (1985): 429–53; Pielke, *The Honest Broker*.

22. Funtowicz and Ravetz, *Uncertainty and Quality in Science for Policy*; Morgan and Henrion, *Uncertainty: A Guide to Dealing with Uncertainty in Quantitative Risk and Policy Analysis*; Rowe, "Understanding Uncertainty."
23. Goodnight, "The Personal, Technical, and Public Spheres of Argumentation," 252.
24. Quoted in Stephen C. Zehr, "Public Representation of Scientific Uncertainty about Global Climate Change," *Public Understanding of Science* 9, no. 2 (2000): 96. See also on the ambiguity of "boundary work" in science Thomas F. Gieryn, "Boundary-Work and the Demarcation of Science from Non-Science: Strains and Interests in Professional Ideologies of Scientists," *American Sociological Review* 48, no. 6 (1983): 85–103.
25. Shackley and Wynne, "Representing Uncertainty in Global Climate Change Science and Policy."
26. For illustrations of achievable provisional political certainty on climate change, cf. the British and German legislative responses to science advice in the 1990's and early 2000's in Jasanoff, "Cosmopolitan Knowledge: Climate Science and Global Civic Epistemology"; Weingart, Engels, and Pansegrau, "Risks of Communication: Discoures on Climate Change in Science, Politics, and the Mass Media." For arguments about the unattainability of scientific certainty in climate change debates, see Hulme, *Why We Disagree About Climate Change*, 106; Pielke, *The Climate Fix*, 201.
27. Hulme, *Why We Disagree about Climate Change*, 84; Dale Jamieson, "Prediction in Society," in *Prediction: Science, Decision Making, and the Future of Nature*, ed. Daniel Sarewitz, Roger A. Pielke Jr., and Radford Byerly Jr. (Washington, DC: Island Books, 2000), 322.
28. Pielke, *The Climate Fix*, 201–2.
29. Shackley and Wynne, "Representing Uncertainty in Global Climate Change Science and Policy: Boundary-Ordering Devices and Authority," 281.
30. Intergovernmental Panel on Climate Change, "The Preparation of IPCC Reports," accessed January 6, 2012, 2012, http://www.ipcc.ch/pdf/ipcc-principles/IPCC%20Procedures.pdf.
31. Intergovernmental Panel on Climate Change, "Guidance Notes for Lead Authors of the IPCC Fourth Assessment Report on Addressing Uncertainties," http://www.ipcc.ch/pdf/assessment-report/ar4/wg1/ar4-uncertaintyguidancenote.pdf; Martin Manning et al., "Workshop Report" (paper presented at the IPCC Workshop on Describing Scientific Uncertainties in Climate Change to Support Analysis of Risk and of Options, National University of Ireland, Maynooth, Co. Kildare, Ireland, May 11–13, 2004), 37.
32. Weiss, "Expressing Scientific Uncertainty," 26.
33. Moss and Schneider, "Uncertainties in the IPCC TAR: Recommendations to Lead Authors for More Consistent Assessment and Reporting."

34. James Titus and Vijay Narayanan, "The Risk of Sea Level Rise: A Delphic Monte Carlo Analysis in Which Twenty Researchers Specify Subjective Probability Distributions for Model Coefficients within Their Respective Areas of Expertise," *Climatic Change 33,* no. 2 (1996); T. Webler et al., "A Novel Approach to Reducing Uncertainty: The Group Delphi," *Technological Forecasting and Social Change 39,* no. 3 (1991): 151–212.
35. Weiss, "Expressing Scientific Uncertainty," 26.
36. Manning et al., "Workshop Report," 103.
37. Carol L. Rogers, "The Importance of Understanding Audiences," in *Communicating Uncertainty: Media Coverage of New and Controversial Science.,* ed. Sharon M. Friedman, Sharon Dunwoody, and Carol L. Rogers (Mahwah, NJ: Lawrence Erlbaum Associates, 1999), 195.
38. Stephen Schneider, "The Dangers of Climate Change, Stephen Schneider (Part 1 of 6), Radio (Transcript)," Global Public Media, last modified June 23, accessed June 16, 2009, http://old.globalpublicmedia.com/transcripts/453. For an inside commentary on the tension the IPCC felt while writing the TAR between elaborating uncertainties and stressing certainties, see John S. Dryzek, Richard B. Norgaard, and David Schlosberg, *The Oxford Handbook of Climate Change and Society* (Oxford, UK: Oxford University Press, 2011), 19, Ebooks Library.
39. Intergovernmental Panel on Climate Change, "Guidance Notes for Lead Authors of the IPCC Fourth Assessment Report on Addressing Uncertainties," 2.
40. Ibid., 3.
41. Pielke, *The Climate Fix,* 201–2.
42. Intergovernmental Panel on Climate Change, "Summary for Policymakers," in *Climate Change 2007: Synthesis Report.,* ed. Core Writing Team, R. K. Pachauri, and A. Reisinger (Cambridge, UK and New York: Cambridge University Press, 2007), 8, http://www.ipcc.ch/pdf/assessment-report/ar4/syr/ar4_syr_spm.pdf.
43. Intergovernmental Panel on Climate Change. "Summary for Policymakers." In Climate Change 2007: The Physical Science Basis. Contribution of Working Group I to the Fourth Assessment Report of the Intergovernmental Panel on Climate Change, edited by S. Solomon, D. Qin, M. Manning, Z. Chen, M. Marquis, K.B. Averyt, M.Tignor and H.L. Miller (New York: Cambridge University Press, 2007), 1, http://www.ipcc.ch/pdf/assessment-report/ar4/wg1/ar4-wg1-spm.pdf.
44. Carolyn Miller makes a low/high, is/ought stasis division that is nearly identical to mine (she excludes the cause/effect stasis). She locates technical evaluations like "very small chance of loss of containment" on the "is" side of the division; see "The Presumptions of Expertise: The Role of Ethos in Risk Analysis." While I disagree on this point—I read these statements as evaluations referencing disciplinary values like efficiency and accuracy—nevertheless, we both conclude that such statements are read as value judgments when they concern high-risk impacts such as nuclear meltdown.

45. Andrew Revkin, "Why 2007 I.P.C.C. Report Lacked 'Embers,'" *Dot Earth (blog)*, February 26, 2009, http://dotearth.blogs.nytimes.com/2009/02/26/why-2007-ipcc-report-lacked-embers/.
46. Intergovernmental Panel on Climate Change, *Summary for Policymakers*, in *Synthesis Report*, 19.
47. Fifteen of the IPCC authors went on to publish the updated "burning embers" diagram independently in 2009; see Joel B. Smith et al., "Assessing Dangerous Climate Change through an Update of the Intergovernmental Panel on Climate Change (IPCC) 'Reasons for Concern,'" *Proceedings of the National Academy of Sciences* 106, no. 11 (2009): 4133–37.
48. There are multiple references to the significance of the AR4 with respect to skepticism about scientific consensus on anthropogenic warming on, for example, the blogs climateark.org, climateprogress.org, realclimate.org, and greenfacts.org. For a more centrist example of the same rhetoric, see Oppenheimer et al.'s "Climate Change: The Limits of Consensus," *Science* 14 (September 2007): 1505–6; this article's title belies its argument—that the unprecedented consensus reflected in the AR4 may mislead policymakers into downplaying or ignoring some of the more dire possible outcomes of anthropogenic warming excluded from the IPCC consensus.
49. Stanford University, "Large Majority of Americans Support Goverment Solutions to Address Global Warming," *Woods Institute for the Environment* website, http://woods.stanford.edu/research/americans-support-govt-solutions-global-warming.html, accessed January 10, 2012.
50. "IPCC's Artful Bias," *Washington Pest* (blog), February 3, 2007, http://thewashingtonpest.blogspot.com/2007/02/ipccs-artful-bias.html.
51. "IPCC Criticism of the Week," last modified August 1, http://globalwarming-questions.googlepages.com/ipcc, accessed June 15, 2009,
52. Indur Goklany, "How the IPCC Portrayed a Net Positive Impact of Climate Change as a Negative," *Cato @ Liberty* (blog), September 18, 2008, http://www.cato-at-liberty.org/2008/09/18/how-the-ipcc-portrayed-a-net-positive-impact-of-climate-change-as-a-negative/.
53. John McLean of the Science and Public Policy Institute compiled what he claimed was a list of "50 articles that seriously question the credibility and integrity of the IPCC's activities and claims," one of which was his own article claiming the SPM authors were biased because many of them had previously published scientific papers together; see "The IPCC under the Microscope," accessed June 15, 2009, http://mclean.ch/climate/IPCC.htm. Another was a critique of the SPM from the blog *spiked* by James Woudhuysen and Joe Kaplinsky. These authors argued that by digesting and summarizing the AR4, the SPM was unwittingly participating in the fomenting of a "New Scientism" where "doubt, a key ingredient of the scientific method, is now out"; see "A Man-Made Morality Tale," accessed June 16, 2009, http://www.spiked-online.com/index.php?/site/article/2819/.

54. Myanna Lahsen, "Seductive Simulation? Uncertainty Distributions around Climate Models," *Social Studies of Science 35,* no. 6 (2005): 897–99.
55. Schneider, Stephen. "The Dangers of Climate Change, Stephen Schneider (Part 1 of 6), TUC Radio (transcript)," *Global Public Media,* accessed June 16, 2009, http://old.globalpublicmedia.com/transcripts/453.
56. Lahsen, "Seductive Simulation? Uncertainty Distributions around Climate Models," 899.
57. There are a plethora of examples available of these types of critiques on the Internet, but as a representative sample from least to most technical, see P. Gosselin, "Fred Singer at Suppressed Seii Presentation: 1976 to 2000 Warming: 'That's Fake, It Doesn't Exist'" (blog), 2011, http://notrickszone.com/2011/09/03/fred-singer-at-suppressed-seii-presentation-1976-to-2000-warming-thats-fake-it-doesnt-exist/; "Will IPCC Climate Models Make Accurate Predictions?" *TakeOnIt,* accessed January 6, 2012, http://www.takeonit.com/question/69.aspx; Stephen McIntyre and Ross McKitrick, "Corrections to the Mann et al. (1998) Proxy Data Base and Northern Hemispheric Average Temperature Series," *Energy & Environment* 14, no. 6 (2003): 751–71.
58. Don Luymes, "The Rhetoric of Simulation in Forest Design: Some Research Directions," in *Forests and Landscapes: Linking Ecology, Sustainability, and Aesthetics, IUFRO 6* (Wallingford, UK: CABI Publishing, 2001): 191–204; Moss and Schneider, "Uncertainties in the IPCC TAR."
59. Warren Thorngate and Mahin Tavakoli, "Simulation, Rhetoric, and Policy Making," *Simulation & Gaming Online,* February 6, 2009, http://sag.sagepub.com/cgi/rapidpdf/1046878108330539v1.
60. See Oreskes, *Merchants of Doubt,* for several case studies of critics—including the ubiquitous Dr. Singer—who use uncertainty to call the ethos of science advisors into question.
61. Martin Parry, "The IPCC: As Good as It Gets," *BBC News* (November 13, 2007), http://news.bbc.co.uk/2/hi/science/nature/7082088.stm. Notice Parry also uses the more strongly prophetic term "prediction" here instead of climate scientists' generally preferred term, "projection." According to IPCC guidelines published in the TAR and after in an attempt to discipline uncertainty, IPCC writers were supposed to use "projection" to refer to the results of models that incorporated socioeconomic assumptions and thus reflected higher uncertainties; "prediction" was supposed to be avoided except in the case of "probable" outcomes. However, a study by Hans von Storch and Dennis Bray in 2009 found that as many as one in four climate scientists actually use these terms *backward,* with "projection" for "probable" and "prediction" for possible. So, the projection/prediction distinction remains undisciplined in both internal and external arguments about climate models; see "'Prediction' or 'Projection'?" *Science Communication 30,* no. 4 (2009): 534–43.

62. Along similar lines, Aimee Kendall found in her study of the use of computer simulations in legal trials that such models can lend "virtual" support to traditional scientific argument: "Computer Simulations in Computational Science: Mimesis of a Virtual Nature," *Professional Studies Review* 2, no. 1 (2005): 31–47.
63. Bruno Latour, *Science in Action: How to Follow Scientists and Engineers Through Society* (Cambridge, MA: Harvard University Press, 1987); "How to Be Iconophilic in Art, Science, and Religion," in *Picturing Science, Producing Art*, ed. Caroline Jones and Peter Galison (New York: Routledge, 1998), 418–40.
64. Miller, "The Presumptions of Expertise: The Role of Ethos in Risk Analysis"; Carolyn R. Miller, "Expertise and Agency: Transformations of Ethos in Human-Computer Interaction," in *The Ethos of Rhetoric*, ed. Michael Hyde (Columbia: University of South Carolina Press, 2004), 197–218.
65. Lawrence J. Prelli, "Visualizing a Bounded Sea: A Case Study in Rhetorical Taxis," in *Rhetorics of Display*, ed. Lawrence J. Prelli (Columbia: University of South Carolina Press, 2006), 90–120.
66. I have it on authority from an IPCC WG3 author that the error bars and model ranges in AR4 WG1/SYR SPM 5 are equal to or greater than those expressed in TAR WG1I SPM 6, the closest related graphic from the previous report. Since uncertainties are not discussed in the AR4 SPM's, I have no means of independently verifying this claim.
67. Kress and Van Leeuwen, *Reading Images*, 198.
68. Ibid., 166.
69. G. Nigel Gilbert and Michael Mulkay, *Opening Pandora's Box* (Cambridge, UK: Cambridge University Press, 1984), 159–67.
70. Roger Pielke, Jr., "Verification of IPCC Temperature Forecasts 1990, 1995, 2001, and 2007," *Prometheus* (blog), January 14, 2008, http://sciencepolicy.colorado.edu/prometheus/.
71. Jamieson, "Prediction in Society," 319.
72. Gilbert and Mulkay, *Opening Pandora's Box*, 155; Kress and Van Leeuwen, *Reading Images*, 160.
73. Kress and Van Leeuwen, *Reading Images*, 145.
74. "IPCC Criticism of the Week"; David Pratt, "Climate Change Controversies," accessed February 17, 2012, http://davidpratt.info/climatecon2.htm; S. Fred Singer, "Nature, Not Human Activity, Rules the Climate," *Heartland Institute*, http://heartland.org/sites/all/modules/custom/heartland_migration/files/pdfs/22835.pdf. accessed February 17, 2012,
75. Ross McKitrick et. al., "Independent Summary for Policy Makers," last modified March 2, accessed June 15, 2009, http://www.fraserinstitute.org/Commerce.Web/product_files/IndependentSummary5.pdf.
76. Gilbert and Mulkay, *Opening Pandora's Box*, 150.

77. Stanford University, "Large Majority of Americans Support Goverment Solutions to Address Global Warming."
78. Andrew Revkin, "Hacked E-Mail Is New Fodder for Climate Dispute," *DotEarth* (blog), *New York Times* website, November 20, 2009, http://www.nytimes.com/2009/11/21/science/earth/.
79. Cf. for criticisms centering around suppressing opposing views and subverting peer review Steve Watson, "'Climategate': Peer-Review System Was Hijacked by Warming Alarmists," *Infowars* (blog), November 23, 2009, http://www.infowars.net/articles/november2009/231109climategate.htm; Robert Tracinski, "Climategate: The Fix Is In," *RealClearPolitics* (blog), November 24, 2009, http://www.realclearpolitics.com/articles/2009/11/24/the_fix_is_in_99280.html; Chip Knappenberger, "Climate-Gate: Is Peer-Review in Need of Change?" *Science and Public Policy* (blog), December 2, 2009, http://scienceandpublicpolicy.org/commentaries_essays/climategate_peer_review.html.
80. Revkin, "Hacked E-Mail Is New Fodder for Climate Dispute."
81. "Their Charms Proved Irresistible," *Finem Respice,* accessed January 6, 2010, http://finemrespice.com/node/72.
82. Meyer, "You're Absolutely Wrong and I Agree with You," *Climate Skeptic* (blog), November 23, 2009, http://www.climate-skeptic.com/2009/11.
83. Goklany, "How the IPCC Portrayed a Net Positive Impact of Climate Change as a Negative,"(blog); McLean, "The IPCC under the Microscope."; McKitrick, "Independent Summary for Policy Makers." Lest I give the impression that these ethical attacks are lobbed only at the scientists who recommend (or whose work grounds recommendations of) activism against anthropogenic warming, there are myriad examples of exactly the same types of arguments leveled against skeptical scientists. Just a few quick examples to illustrate: Michael Mann refers to McKitrick and McIntyre's (2003) criticisms of the methods behind Mann's famous "hockey stick" graph as "refuted claims . . . by two Canadians (an economist and an oil industry consultant)," obviously implying the researchers' industry connections cast aspersions on their scientific findings: quoted in Gavin Schmidt, "The Discovery of Global Warming' Update—Comment #2," *RealClimate* (blog), July 14, 2006, http://www.realclimate.org/index.php/archives/2006/07/the-discovery-of-global-warming-update/. And *Crooked Timber* lumps McKitrick with "'sceptics' [who] are not, as they claim, fearless seekers after scientific truth, but ideological partisans and paid advocates, presenting dishonest arguments for a predetermined party-line conclusion" (para. 6); see John Quiggin, "The End of the Global Warming Debate," *Crooked Timber* (blog), January 4, 2006, http://crookedtimber.org/2006/01/04/the-end-of-the-global-warming-debate/.

CHAPTER 10

1. Stephen S. Hall, "Scientists on Trial: At Fault?" *Nature* 477 (2011), http://www.nature.com/news/2011/110914/full/477264a.html.
2. Cf. Fahnestock, *"Accommodating Science"*; Nelkin, *Selling Science*; and, Winsten, "Science and the Media."
3. Jasanoff, *The Fifth Branch*, 16.
4. Intergovernmental Panel on Climate Change, "Mandate," *About IPCC*, accessed January 10, 2010, http://www.ipcc.ch/about/index.htm.
5. Jasanoff, "Cosmopolitan Knowledge: Climate Science and Global Civic Epistemology," 138; Weingart, Engels, and Pansegrau, *"Risks of Communication: Discourse on Climate Change in Science, Politics, and the Mass Media."*
6. Zehr, *"Public Representation of Scientific Uncertainty About Global Climate Change,"* 86–87.
7. See Gregory and Miller, *Science in Public*, 120, for the argument that communication scholars have no evidence that if they could fix the known problems with science journalism, it would make a difference in the public understanding of science. See Morris Shamos, *The Myth of Scientific Literacy* (Piscataway, NJ: Rutgers University Press, 1995) for the argument that scientific literacy as currently defined is an unattainable ideal for the lay voter.
8. Rogers, *"The Importance of Understanding Audiences,"* 180–82.
9. For a review of models for incorporating local knowledge and values in policymaking, see Craig Waddell, "Saving the Great Lakes: Public Participation in Environmental Policy," in *Green Culture: Environmental Rhetoric in Contemporary America*, ed. Carl G. Herndl and Stuart C. Brown (Madison, WI: University of Wisconsin, Press, 1996), 141–165. For the two examples cited, see this study by a legal expert Anne M. Hartridge, "Salmon Medicine: Federal Trust, the ESA, and the Trinity River," *Environs: Environmental Law & Policy Journal* 23 (1999): 107–130; also, this study by pediatricians Joshua L. Bonkowsky et al., "Metamizole Use by Latino Immigrants: A Common and Potentially Harmful Home Remedy," *Pediatrics* 109, no. 6 (2002).
10. Marver Bernstein, *Regulating Business by Independent Commission*, (Princeton, NJ: Princeton University Press, 1955).
11. As one senator in *The New Priesthood*, Ralph Lapp's study of science policymaking expressed his frustration with trying to keep abreast of the technical aspects of legislation:

    a. We have water research bills, fish research bills, mineral development research bills, urban psychological study programs, supersonic airplane bills, satellite communication bills, air pollution, water pollution, environmental hazard bills, cancer, heart disease and mental health bills, retardation and epilepsy programs, moose conservation and forestry research bills. Who is to tell the Congress what is redundant, what is necessary, what is duplication, and what is vital? (154)

12. Pielke, *The Honest Broker,* 99.
13. Ibid., 18.
14. Pielke, *The Climate Fix,* 202.
15. *The Honest Broker,* 52.
16. See Cornelia Dean, "Groups Call for Scientists to Engage the Body Politic," *New York Times* website, August 8, 2011, http://www.nytimes.com/2011/08/09/science/09emily.html, for a brief discussion of Ben Franklin's list and other efforts to increase the number of science, technology, engineering, and math (STEM) professionals in Congress. STEM workers and medical professionals currently make up about 7% of the House and 1% of the Senate for an aggregate percentage of 6%. The figure of 10% for national STEM and medical professionals comes from the 2005–2009 community survey figures found at factfinder.census.gov. If medical professionals are subtracted, STEM workers in the U.S. make up about 3% of the employed adult population.
17. Ibid.
18. Brooks, "The Scientific Adviser"; Price, "The Scientific Establishment"; Wood, "Scientists and Politics: The Rise of an Apolitical Elite"; Lapp, *The New Priesthood;* Latour, "How to Be Iconophilic in Art, Science, and Religion"; Brian Appleyard, *Understanding the Present: An Alternative History of Science* (New York: Doubleday, 2003). Also, see Joseph Ben-David's comparative study of science policymaking, on the basis of which he argues that "a permanent attempt to extend scientific thinking to human and social affairs" must characterize any society which intends to pursue science "without recurrent moral crises"; *The Scientist's Role in Society: A Comparative Study* (Chicago, IL: University of Chicago Press, 1971, 1984), 184.
19. Ferris, *The Science of Liberty,* 290–91.
20. Shamos, *The Myth of Scientific Literacy;* Patrick Sturgis and Nick Allum, "Science in Society: Re-Evaluating the Deficit Model of Public Attitudes," *Public Understanding of Science 13,* no. 1 (2004): 55–74. See George DeBoer, "Scientific Literacy: Another Look at Its Historical and Contemporary Meanings and Its Relationship to Science Education Reform," *Journal of Research in Science Teaching 37,* no. 6 (2000): 582–601, for a review of definitions of scientific literacy and arguments about its utility as an empirical education research variable.
21. Miller's test, the only one of its kind, asks questions about basic scientific constructs like lasers or DNA as well as questions about traditional scientific-logic processes. It has been administered to samples as large as 2,000 in as many as 40 countries, mostly the United States and the European Union. See for more information about Civic Scientific Literacy, "The Measurement of Civic Scientific Literacy"; also, Jon D. Miller, "Public Understanding of Science" (paper presented at the 93rd Annual Ecological Society of America Annual Meeting, Milwaukee, WI, August 3–8, 2008). If 10% of the US adult population

was CS literate in 1988, and 25% was CS literate in 2008, this means CSL is currently doubling every 16 years.

22. Estimates range, but Kevin Kelly, the editor of *Wired*, and Hal Varian from Google estimate that scientific publications double every 15 years. But that figure is for publications, not knowledge, which undoubtedly increases more rapidly: "The Expansion of Ignorance," *The Technium* (blog), 2008, http://www.kk.org/thetechnium/archives/2008/10/the_expansion_o.php; and "The Speed of Information," *The Technium* (blog), 2008, http://www.kk.org/thetechnium/archives/2006/02/the_speed_of_in.php. One intriguing study of the rate of knowledge growth in yeast biology research over the last 30 years has estimated that knowledge is doubling in that field about every 1.5 to 2 years: see Xionglei He and Jianzhi Zhang, "On the Growth of Scientific Knowledge: Yeast Biology as a Case Study," *PLoS Comput Biol* 5, no. 3 (2009): e1000320.
23. For some high-profile examples of this topos, see Edwards, *A Vast Machine*, 436; Hulme, *Why We Disagree about Climate Change*, 82. I have also heard these arguments repeated informally by numerous climate scientists.
24. Manning et al., "Workshop Report," 103.
25. Besel, "Communicating Climate Change," 193–96; Weiss, "Expressing Scientific Uncertainty," 27. Also, climate scientist Armin Haas of the Potsdam Institute of Climate Impact Research made a related argument about the potential of economic metaphors in a personal conversation.
26. Jamieson, "Prediction in Society," 320.
27. For some examples of local environmental policymaking that has balanced local stakeholder values with scientific advice, see Brian W. Eisenhauer and Brian Nicholson, "Using Stakeholders' Views: A Social Science Methodology for the Inclusive Design of Environmental Communications," *Applied Environmental Education & Communication* 4, no. 1 (2005): 19–30; Hartridge, "Salmon Medicine: Federal Trust, the ESA, and the Trinity River"; Waddell, "Saving the Great Lakes: Public Participation in Environmental Policy."
28. *Hulme, Why We Disagree about Climate Change*, 105–7; Steve Rayner, "Prediction and Other Approaches to Climate Change Policy," *in Prediction: Science, Decision Making, and the Future of Nature*, ed. Daniel Sarewitz, Roger A. Pielke, Jr., and Radford Byerly, Jr. (Washington, DC: Island Books, 2000), 285.
29. Hulme, *Why We Disagree about Climate Change*, 309.
30. For the term "civic epistemology," meaning a polity's preferred method for establishing truths, see Jasanoff, "Cosmopolitan Knowledge: Climate Science and Global Civic Epistemology."

INDEX

The manufacturer's authorised representative in the EU for product safety is Oxford University Press España S.A. of El Parque Empresarial San Fernando de Henares, Avenida de Castilla, 2 - 28830 Madrid (www.oup.es/en or product.safety@oup.com). OUP España S.A. also acts as importer into Spain of products made by the manufacturer.
Printed and bound by CPI Group (UK) Ltd, Croydon, CR0 4YY

14/09/2026

02216284-0011